River-Sand Mining

China Studies

PUBLISHED FOR THE INSTITUTE FOR CHINESE STUDIES, UNIVERSITY OF OXFORD

Series Editor

Rana Mitter, *University of Oxford*

VOLUME 45

The titles published in this series are listed at *brill.com/chs*

River-Sand Mining

An Ethnography of Resource Conflict in China

By

Qian Zhu

BRILL

LEIDEN | BOSTON

Cover illustration: Nanning City, photo taken by author.
2014, Grant Application for a research stay abroad within the IPID4all a.r.t.e.s.international program University of Cologne.
2015, DAAD-Network remapping the Global South at the Global South Studies Center (GSSC) University of Cologne.
2018, A proofreading grant from the a.r.t.e.s Graduate School University of Cologne.

The Library of Congress Cataloging-in-Publication Data is available online at https://catalog.loc.gov
LC record available at https://lccn.loc.gov/2021051108

Typeface for the Latin, Greek, and Cyrillic scripts: "Brill". See and download: brill.com/brill-typeface.

ISSN 1570-1344
ISBN 978-90-04-50590-2 (hardback)
ISBN 978-90-04-50591-9 (e-book)

To my grandmother

Contents

Preface

China, like many countries, struggles with the conflict between economic development and a sustainable use of natural resources. This book examines one source of conflict, namely the overexploitation of river sand for the purpose of regional development in rural southwest China and its impact on local communities. By applying a political ecology approach to the management of river sand, one of the natural resources in highest demand, the book offers an insightful analysis of the complex governance structures that regulate river-sand mining on the one hand, and on the other hand the catastrophic over-extraction of river sand and its social and ecological impacts. In particular, the author focuses on the illegal mining of river sand in an area home to the Zhuang, the largest officially recognized ethnic minority group in China.

As a member of the Zhuang and a local resident, the author was able to engage with a diverse group of relevant stakeholders over the course of her fieldwork, including local government officials and party cadres; villagers and their representatives; unlicensed sand miners, wholesalers, truck and ship owners; and workers in the mining industry. Drawing on rich ethnographic material, the author captures the perspectives of different actors and the power dynamics in sand mining. She also discusses the contradictory laws that regulate the business and the fragmentation of authority between and within different levels of government, showing how these factors are increasing rather than decreasing the number of stakeholders and conflicting interests when it comes to sand management.

This book highlights the exclusion of riparian residents from decision-making processes pertaining to their land. Their disempowerment has allowed the practice of illegal sand mining to continue for decades and contributed to environmental injustice.

Since the late twentieth century, southwest China has witnessed the large-scale commodification of river sand, most of which is obtained illegally. The commodification results from a high demand for river sand to support large-scale infrastructure developments such as the construction of roads, bridges, houses, and dams, leading to a dramatic increase in sand prices in both rural and urban areas. The objective of illegal sand-mining activities has been purely economic, with little concern for its effects on the environment, including riverbanks, water quality, adjacent farmlands, and aquatic life. Indeed, sand mining has enormous social, economic, and ecological implications, as discussed throughout this book.

The book focuses on the market orientation of river-sand mining and its sociocultural and ecological consequences in rural Zhuang communities around the Maoling River—the largest river in Qinzhou City. It also introduces the diverse actors involved, including government officials, riparian Zhuang communities, and legal and illegal miners. The multiplicity of actors relates to the growing complexity of institutions and policies at various levels, which contributes to local-level disputes, conflicts, and the mismanagement of sand resources. This book takes on a political ecology perspective to explore resource conflicts and sand exploitation, addressing the intertwined issues of institutions, power, contention, scale, justice, and resistance.

Rampant river-sand extraction takes a severe toll on the rural Zhuang villages, affecting villagers' property, territory, customs, agricultural production, and daily activities. River sand plays a vital role in the river ecosystem. The over-appropriation of river sand has led to faunal destruction (i.e., the loss of fish species), water pollution, and the collapse of farmland as a result of erosion. In order to reduce their loss, local villagers have practiced various forms of resistance against illegal river-sand mining. This book uses detailed empirical research, conducted in six riparian Zhuang villages, to explore the causes and consequences of illegal river-sand mining in southwest China.

Acknowledgments

I am deeply grateful to the anonymous reviewer who helped me to broaden my theoretical discussion by suggesting some crucial works. My engagement with the recommended texts enabled me to place my research within a wider theoretical framework. I greatly appreciate the reviewer's insightful comments and suggestions.

It would have been impossible to finish writing this book without great assistance from many helpful professors, colleagues, friends, and family members, among others.

I would like to begin by thanking my first thesis advisor, Prof. Dr. Michael Bollig, who has much experience in supervising masters and Ph.D. students. I am very grateful for his patience in allowing me to take my time to improve my knowledge of anthropology, and for agreeing to serve as my supervisor in the Culture and Environment in Africa (CEA) masters program, which I completed in summer 2018. Importantly, he let me freely choose my own research topics until I eventually settled on the topic that is at the heart of this dissertation. Furthermore, he taught me the major theories, e.g., political ecology and historical ecology, that I have applied in my writings. He also helped me greatly in clarifying what the book is about. Following his advice, I paid more attention to the Zhuang people, who share my own ethnic identity, and thus he helped to steer me toward studying the relationship between human beings and their environment, which proved to be a productive approach.

I would also like to thank my second supervisor—Prof. Dr. Michaela Pelican. I am very lucky to have joined her colloquium and many other courses. I have greatly benefited from her advice to look for a gap in the existing research and to bridge it by doing my own research and writing my own ethnography. Furthermore, her review of my chapters has given me many crucial insights that have helped me to improve this book.

I greatly appreciate the teaching and help of Prof. Dr. Gerrit Dimmendaal and Dr. Clemens Greiner, who have given me insightful advice during my studies. My kind regards also go out to the Global South Studies Center (GSSC) and the a.r.t.e.s. Graduate School of the University of Cologne for supporting me with an office, assistance, and funds for conducting my research.

I am deeply in debt to Prof. Dr. Daming Zhou and Prof. Dr. Tan Chee-Beng in the School of Sociology and Anthropology at Sun Yat-sen University. I was so lucky to join their classes when I visited Sun Yat-sen University in 2015. Since then, they have consistently supported me in my academic research.

There are other people who have supported me in diverse ways. First, I appreciate Dr. Gerda Kuiper, who helped me during the process of proofing and publishing this book. I also thank Dr. Christiane Naumann and Dr. Eric Mutisya Kioko, who read the first draft of my dissertation and gave me very detailed feedback. I also thank Bright Masocha and other colleagues who read some chapters and helped them to make more sense. Indeed, I cannot mention all the names of those who helped me here, but I will remember their kindness and assistance.

Last, but not least, I am grateful to the people of Qinzhou City for helping me conduct my fieldwork. Both villagers and the officials agreed to give interviews, fill in questionnaires, and participate in focus-group discussions, all of which greatly supported this research. Without their generous cooperation, this thesis would have been impossible. Finally, I thank my family, relatives, and friends both at home and in Germany for their love and support.

Illustrations

Maps

Figures

Tables

Notes on the Text

Abbreviations

CCP	China Communist Party
CCTV	China Central Television
FYP	Five-Year Plan
GDP	Gross Domestic Product
GZAR	Guangxi Zhuang Autonomous Region
HRS	Household Responsibility System
NGOs	Non-Governmental Organizations
NPC	National People's Congress
PPPs	Public–Private Partnerships
PRH	Public Rental Houses
PRC	People's Republic of China
RMB	Renminbi (ISO-4217-Code: CNY), the official currency of the People's Republic of China
UNEP	United Nations Environment Programme

Units

The *mu* is the Chinese unit of area; 1 *mu* = 666.7 m^2.

The yuan (CNY, also RMB) is the Chinese unit of currency; 1 euro = 7.9 yuan (August 2017).

CHAPTER 1

Introduction

This book is about the illegal mining of river sand in six Zhuang villages in southwest China. It discusses the actors involved, the complex policy environment in which this illicit enterprise thrives, its negative consequences, and local resistance to it. The book adopts a political ecology perspective to understand how power relations at both the formal/state and informal/local levels play out and contribute to the increasingly dangerous trade in river sand—a source of ecological, sociocultural, and economic harm. River sand has emerged as the latest natural resource to be rapidly mined by diverse parties. This phenomenon of rapid extraction, of course, is closely related to the political, environmental, hydrological, socioeconomic, and cultural situation of the people involved in and affected by the mining (Krausmann et al. 2009; Krause, Diesing, and Arlt 2010; Maya et al. 2012). The findings show that economic and social developments in the villages have been mainly driven by the state, accompanied by the commercialization of natural resources and rural labor.

China has many minority groups with diverse indigenous ecological knowledges. This book looks at illegal river-sand mining taking place in such a minority region. It is based on long-term ethnographic research among members of the largest minority group in China—the Zhuang minority in the Guangxi Zhuang Autonomous Region (GZAR). It describes the implications of sand mining in six Zhuang villages located in two different administrative towns.

The illegal mining of sand has resulted in overexploitation and a sand crisis, revealing the complexity of river-sand management by both formal and informal institutions. In their article "A Looming Tragedy of the Sand Commons," Torres, Brandt, et al. (2017, 970) state: "In most regions, sand is a common-pool resource, i.e., a resource that is open to all because access can be limited only at high cost." River sand is likewise considered a common-pool resource in the Zhuang villages along the rivers, and in the past it was managed by informal institutions within the village (Zhu 2018). However, the lineage system was destroyed by the state. As Siu (2016, 98) notes,

> As early as the land reform from 1950 to 1952, landed estates on which ancestral halls and temples depended to finance activities were confiscated. … During the Cultural Revolution, young students were mobilized to attack "the four olds," or traditional customs and habits; classic texts and genealogies were burned, and the images of deities were smashed.

 | DOI:10.1163/9789004505919_002

In addition to the destruction of ancestral halls, the weakening of lineage leadership, and the prohibition of old rituals for births, funerals, and weddings, informal institutions in the Zhuang villages have been eroded. The villages lack effective informal institutions to manage river sand as a common-pool resource and to mitigate conflicts over sand mining. Furthermore, the government withdrew its involvement on the village level in the 1980s, creating a power vacuum (Siu 1989a, 2016; Faure and Siu 1995). Some scholars have noted the subsequent revival of religious movements and rituals in China (Lai 2003; F. Yang 2011; Siu 2016), but these religious institutions are powerless against illegal river-sand mining and the forces behind the sand industry.

The overexploitation of sand has impacted local ecological knowledge as expressed in the customs, beliefs, and social relationships of the Zhuang people. Aswani, Lemahieu, and Sauer (2018, 1–3) state:

> The value of local ecological knowledge (LEK) (also referred to as indigenous and traditional ecological knowledge [IEK or TEK]) to mainstream conservation, began in the early 1990s when various researchers showed the cost-effectiveness and fairness to indigenous peoples' territorial rights of integrating LEK into development and conservation projects. ... LEK is site specific knowledge held by a group of people about the various components of their environment and that may integrate both scientific and practical knowledge. ... We consider LEK in its most inclusive definition of ecological knowledge, including local, indigenous, traditional, and rural and which may be held by an indigenous community and/or local/rural communities from developing or industrialized nations.

Some scholars believe that local and indigenous knowledge may contribute to solving the crisis of natural resource management and environmental degradation in rural areas inhabited by ethnic minorities (Anderson 1996; Senanayake 2006; Von Glasenapp and Thornton 2011; Duara 2015). However, these works have not explored how in the ensuing processes of negotiation, compromise, dispute, and conflict, the territory has been changed physically and discursively. My findings reflect that any local ecological knowledge remains powerless without the support of solid customs and related informal institutions. In China's top-down political system, which has been called "authoritarian" (Landry 2008), the room for informal institutions to govern common resources is decreasing, and attempts to govern by such institutions are even considered illegal.

Formal institutions seem to be much more capable of acting against illegal river-sand mining in China. Other scholars argue that authoritarianism has played a key role in environmental management in China (Li and Shapiro 2020; Beeson 2010, 2016; Gilley 2012). Gilley (2012, 288) defines authoritarian environmentalism

> as a public policy model that concentrates authority in a few executive agencies manned by capable and uncorrupt elites seeking to improve environmental outcomes. Public participation is limited to a narrow cadre of scientific and technocratic elites while others are expected to participate only in state-led mobilisation for the purposes of implementation.

Beeson (2016, 1–2) argues: "One of the most noteworthy features of China's response to climate change is the fact that it is being overseen by an authoritarian government that has a greater capacity to implement potentially difficult reforms." Li and Shapiro (2020, 10) point out that

> China exemplifies a model of state-led, authoritarian environmentalism which concentrates political, economic, and discursive power within the parameters of the state under the centralized leadership of the Communist Party. … By simultaneously expanding the regulatory scope of the state to encompass a growing range of environmental issues and co-opting non-state actors into the state's environmental agenda, the Chinese state goes green.

This book engages with the concept of authoritarian environmental management through a detailed analysis of state laws and policies on river-sand mining in China. The findings show that overlapping laws have caused a fragmentation of enforcement among government departments on both the vertical and horizontal levels. The poor outcome of the most recent policy against illegal river-sand mining challenges the strategy of top-down environmental management. Exclusion of local stakeholders in policy-making, limitation of public participation, and failure to respond to petitions from citizens constrain the effectiveness of state laws and policies.

In short, both informal and formal channels to stop illegal sand mining seem to be weak and ineffective, even though citizens are aware of the consequences of sand mining. The struggles between state policies, local communities, and market forces are explored in detail in this book to illustrate the actual sociocultural and ecological transformation induced by sand mining in southwest rural China. Recently, many foreign scholars have explored China's environmental management from a political perspective. This book builds on this work but also follows ethnography's principle of holistic writing. It analyzes multiple perspectives on China's sand mining, including cultural, social, and economic perspectives as well as the future direction of China. A multi-perspective analysis contributes to a more thorough understanding of environmental issues in today's China and their global implications.

According to Jensen and Bateman (1979), "sand includes almost any comminuted rock or mineral, but technically it is restricted to quartz sand with minor impurities of feldspar, mica, and iron oxides" (quoted in Sreebha 2008, 2). Sand can be classified according to three different aspects: its origin, its composition, and its grain size (Koirala and Joshi 2017, 3–4). In China, sand is categorized based on its source, for instance, river sand, sea sand, desert sand, and pit sand. According to the 2007 Standard for Engineering Classification of Soil (GB/T 50145-2007) in China, classification methods for sand based on the average diameter of sand grains (between 0.075 mm and 2 mm, or larger than 2 mm) are used for engineering and construction purposes (Zhu et al. 2015).

Sand is used for many purposes, such as construction, glass production, agriculture (crops such as watermelon, peaches, and peanuts prefer sandy soil), and land reclamation projects (Koirala and Joshi 2017, 3; Subramanian 2017; Torres, Brandt, et al. 2017). Sing (2004–2005, 121) reports, "Sand is an important construction material of natural origin. Mixed with cement and lime, millions of tons of sand are used every month for construction as mortar, plasters and concrete" (quoted in Koirala and Joshi 2017, 2). As far as sand used for construction is concerned, the natural properties of river sand determine its value. The salinity of sea sand means that it contains chlorine ions, which will corrode steel and render structures built of it susceptible to erosion (e.g., Hong 2002, 2004). Desert sand is too small and too round-grained to bind well and may contain alkalis (Bagnold 1954, Wang and Li 2014, Zhu, et al. 2015). River sand has neither of these flaws and has thus become a "golden material" in the construction market. Sand for construction is mainly mined in rivers, because ocean sand is less suitable. As a result, river sand is eagerly extracted by sand miners. Jordan (1980, 161) stated:

> Sand found in land deposits is known as "pit sand." Such grains are generally irregular, shape and angular. Sand carried by water, such as found along banks of rivers or lakes, is known as "river sand," such grains are generally rounded and smooth, due to the action of water. Both types of sand are suitable for cement work, so long as they are well-graded and clean.

Koirala and Joshi (2017, 3) point out that

> people do not wonder about the origin of sand. Thousands of years need to pass for rocky material to finally turn into sand or clay. The sand that

> eroded from sandstone rocks [was] deposited as a beach, dune or desert. After millions of years, sandstone rocks turned into sandstone cliffs and eventually eroded for the second time.

River sand is a finite resource, an inseparable and important part of rivers, and plays a vital role in the river ecosystem (Sreebha 2008; Wu 2008). River-sand dredging has a severe environmental impact (Collins and Dunne 1986, 1987; Collins 1991; Kondolf 1994a; Kanehl and Lyons 1992; Harvey and Smith 1998; Klingeman et al. 1999; Soman and Baji 1999; Padmalal et al. 2005, 2008; Jia et al. 2007; Wu 2008; Ladson and Judd 2014).

Wu (2008, 110–111) observes that there are four types of sand mining in China:

1. Sand mining to supply the market for construction materials;
2. Sand mining for fill to be used in construction foundations or for land reclamation projects;
3. Sand mining for public welfare purposes, e.g., for reinforcement of dams and riverbanks, river regulation, and waterway dredging;
4. Sand mining for mineral extraction.

This book focuses on the first category, sand mining for constructing infrastructure and houses. In particular, it focuses on illegal river-sand mining activities (unlicensed mining) and their governance since the 1980s. Illegal river-sand mining and governance reflect the more general situation of resource exploitation and regulation in the country with the world's second-largest economy—China—which is an emerging player in the marketization of natural resources, in this case, river sand. This book—based on thirteen months of ethnographic fieldwork—reveals the scarcity of river sand and asserts that it is one of the main resources supporting China's development, effectively transforming China's economic model from an export-led model to a domestic consumer-led model.

1 A Brief History of River-Sand Mining: A National Perspective

China has an abundance of rivers and lakes, and the phenomenon of (illegal) sand mining has a long history, as illustrated by the case of the Yangtze River. With a length of 6,300 km, the Yangtze is the third-longest river in the world (NBS 2001, 5–7). According to the website of the Changjiang (Yangtze River) Water Resources Commission (CJW), the Yangtze is a key river in China's river system and was governed directly by the state council even before the People's

Republic of China was established in 1949. The central government established the Yangtze River Commission to govern the river during the administration of the Republic of China (1912–1949). Since 1949, the Ministry of Water Resources of the People's Republic of China has become the major authority responsible for governing the river. To effectively implement river management, the central government decided to follow the path taken by previous administrations and therefore took over the Yangtze River Water Conservancy Commission, beginning in February 1950, and subordinated it to the Ministry of Water Resources.[1]

According to Wu (2008), illegal sand mining became a huge issue on the Yangtze River in the beginning of the 1970s. In concert with economic development, the massive demand for river sand increased rapidly, triggering sand-mining activities on the middle and lower reaches of the Yangtze River. The sand-mining boom occurred mainly due to the economic development of Shanghai and of Jiangsu Province in this period, as well as the construction of various enormous dams, such as the Gezhou Dam (opened in May 1971) and Three Gorges Dam (opened in December 1994), which demanded a great amount of river sand.

The greater demand for sand also changed the governance targets for the Yangtze River. Governance previously concentrated on water traffic, flood control, the effective implementation of national projects, and the effective management of water resources. Combating illegal river-sand-related activities became the major target of the Yangtze River Commission in the 1970s and reached a peak in the 1990s (Wu 2008, 1). By that time, the institutions and policies responsible for sand mining had been rendered obsolete by current mining processes, leading to a "vacuum" in sand management (Wu 2008, 2). In 1991, the amount of sand excavated from the Yangtze River totaled about 30 million cubic meters (Wu 2008).

Wu (2008, 3–4) regards "rampant illegal sand mining and visible riverbank collapse, soil erosion, riverbed destruction, water pollution, the disturbance and endangerment of the reproductive habitat of rare dolphins, and social unrest in the regions around the river as evidence of the management crisis."

---

1 Similar management systems govern six additional rivers and one lake in China. These include the Water Conservancy Commission of Huanghe (Yellow) River, the Water Conservancy Commission of Pearl River, the Water Conservancy Commission of Songliao River, the Water Conservancy Commission of Huai River, the Water Conservancy Commission of Hai River, and the Administrative Bureau of Taihu Lake. They are directly governed by the Ministry of Water Resources of the People's Republic of China. All of these rivers and the lake have problems with sand-mining management.

Indeed, a huge flood in 1998 forced the government to take action to prohibit any river-sand mining.

Facing the extreme chaos of river-sand mining, the state council established the Regulations on Sand Mining Management for the Yangtze River in 2001, which aim to govern sand mining in the river and set an example for the management of sand-mining in other rivers (Wu 2008). The Yangtze River has its own special institution in terms of sand-mining management. The regulations govern the waterway from Yibin City in Sichuan Province to Hubei Province, Hunan Province, Jiangxi Province, Anhui Province, Jiangsu Province, Chongqing municipality and, by extension, to Shanghai municipality—a total of six provinces and two province-level municipalities. The Regulations on Sand-Mining Management on the Yangtze River give more power to the Ministry of Water Resources and its subordinate authority and to the Yangtze River Water Conservancy Commission, two major organizations governing the Yangtze River's sand.

In the last two decades, China has established a series of national policies on sand management. The Ministry of Water Resources held a national conference on sand management at the Yangtze River in 2004. The deputy minister, Mr. Chen, gave a public talk at the conference and announced that sand governance had made great progress. He said that new sand policies, institutions, and departments concerning sand management had been established at all levels in China since 2002, and that the ministry had built infrastructure to improve river-sand management. More than 150 agencies (collectives) were entrusted with the task of sand-mining management and the enforcement of related laws (Chen 2004).

Although the Ministry of Water Resources celebrated the "positive outcome" of the first coherent sand-management strategy, the problems of illegal sand overexploitation in Chinese rivers are getting worse, including issues such as inadequate administrative supervision, the low cost of illegal mining, corruption, inefficient management, and rampant illegal sand gathering. Illegal sand-mining activities have occurred in major lakes, rivers, and even oceans. In fact, due to the huge demand for river sand, the state's "efforts" to enforce sand-mining governance were not enough. In 2009, the Ministry of Water Resources and the Ministry of Transport signed the "Strengthening of the Yangtze River Sand Mining Management Cooperation Memorandum," paying full attention to functional advantages and working together in order to rigorously proceed against illegal sand mining and to maintain control over sand mining in the Yangtze River, thus promoting a sustainable situation. In 2015, the same two ministries further called for the implementation of the responsibility system for sand-excavation management along the Yangtze River and the implementation of the accountability system (MWR et al. 2015). The cooperating ministries emphasized the need to scientifically plan sand excavation along the Yangtze River, and they issued a draft for the governance of sand excavation on the upper reaches of the Yangtze River (2015–2019), as well as implementing clearly planned management requirements and measurements. At the same time, they organized the preparation of the Yangtze River's middle and lower reaches and created a flow plan for the river channel (2016–2020) (MWR and MTPRC 2016).

In August 2015, the Ministry of Water Resources, the Ministry of Land and Resources, and the Ministry of Transport jointly released an official document titled the "Further Strengthening of River Sand Management Notice" that called for a clear division of responsibilities, strengthening the management of sand mining, and combating illegal sand appropriation. Despite this document, three cases of illegal sand mining were reported on the "Economic 30 Minutes" program on CCTV between January 21 and 23, 2016. On three consecutive days, the program reported sand-management failures in Henan Province, Jiangxi Province, and Hainan Province, where sand had been illegally extracted. The reports exposed the chaos within sand management in these three provinces, from the auction of sand-mining licenses to issues in law enforcement and supervision, contrasting the current situation with the "Further Strengthening of River Sand Management" notice that the three ministries had jointly issued in 2015.

Increasing environmental protests and complaints concerning issues such as the shortage of clean water, arable land, and raw natural resources forced the government to establish further national institutions and policies for constructing a harmonious society and a sustainable economy, including the implementation of the River Chief System (*hezhangzhi*河长制) in 2016 (Y. Zhu 2016). The River Chief System is a political reform that stresses that top government leaders from the provincial level down to the township level should be the "chiefs of the river," who are in charge of and responsible for the river. The government set out to build an administrative system as well as an assessment system to evaluate the effectiveness of the river chiefs' management. In all likelihood, a top official who was held accountable as a river chief would be risking his or her political career if he or she failed to manage a river effectively and sustainably. The aim of the central government in implementing the River Chief System across the country is to solve the environmental crises of rivers and to realize the sustainable use of water, sand, and fish species in the rivers. One of the major targets of this system is to "resolutely crack down on illegal activities involving rivers and lakes and to resolutely clean up illegal sand mining activities" (GOCCPCC and GOSC 2017).[2]

Significantly, ecological conservation or the "construction of ecological civilization" (*shengtai wenming jianshe* 生态文明建设), an issue that was raised in the 18th National Congress of the Communist Party of China in 2012, has recently become a national policy goal in China. Under the latest political reforms, more and more provinces have taken action to reconstruct their environmental governance and resource conservation policies. On June 19, 2018, the "Notice of the General Office of the Ministry of Water Resources on Launching the National River and Lake Sand Mining Special Remediation Action" was issued to all provinces and authorities in China. The notice states:

> Illegal sand mining seriously affects the stability of the river and threatens flood safety, navigation safety, and ecological security. According to the requirements of the central government for the full implementation of the River Chief System, the river chiefs and lake chiefs at all levels are responsible for leading the organization to clean up and rectify the outstanding problems such as illegal sand mining, and to further strengthen the management of sand mining in rivers and lakes, severely crack down on illegal sand mining, and earnestly safeguard the healthy life of rivers and lakes. After research, on June 20, 2018, the Ministry of Water Resources organized a six-month river and lake mining action nationwide.
>
> MWRPRC 2018a; translated by the author

As a consequence of this governmental policy, illegal sand mining in rivers, lakes, and oceans has been stopped in many provinces. Some local policies were drafted in order to effectively implement the notice and related policies. However, there are still problems that need to be dealt with in order to further implement the state's policy. Lack of clarity regarding the boundaries of rivers and lakes has persisted despite a notice issued by the Ministry of Water Resources on December 20, 2018. The "Notice of the Ministry of

---

2 The River Chief System involves an enormous amount of personnel: "More than 300,000 river chiefs in a four-tier system from provincial to township levels have been appointed in the 31 provincial regions in the Chinese mainland, which means the River Chief System had been established across the country half a year earlier than planned, said E Jingping, minister of water resources, in a news conference on Tuesday. He also said 402 of the river chiefs are provincial-level officials and 59 of them are heads of their provinces. All but two of the provincial regions have also appointed 760,000 river chiefs at the village level" (Liqiang Hou 2018).

Water Resources on Accelerating the Delimitation of the Management Scope of Rivers and Lakes" (MWRPRC 2018b) required related government departments to clarify the boundaries of national rivers and lakes by the end of 2020. This notice directly pointed out that "clarifying the boundaries of rivers and lakes is the foundation for strengthening the management of rivers and lakes. ... This is the mission requirement in order to carry out the River Chief System."

In July 2018, the "Notice on Launching Special Actions for Governing Violations of Sea Sands" was jointly issued by eight ministries (agencies), namely the Ministry of Housing and Urban-Rural Development, the Ministry of Public Security, the Ministry of Natural Resources, the Ministry of Ecology and Environment, the Ministry of Transport, the Ministry of Water Resources, the Ministry of Commerce, and the State Administration of the State of the People's Republic of China. The notice was a reaction to the severe ecological and social consequences of illegal sea-sand mining. Therefore, the eight departments at the national level joined together to severely crack down on the illegal mining, trading, and use of sea sand (MOHURD 2018). In order to avoid the overexploitation of sand and fish species, to prevent water pollution, and to protect biodiversity and ecology, the Chinese government recently adopted a new law: the Yangtze River Protection Law, the first legislation to address a specific river basin. It was adopted at the 24th Meeting of the Standing Committee of the 13th National People's Congress on December 26, 2020, and came into force on March 1, 2021 (Xi 2020). This occurred after the Yangtze River white sturgeon was declared extinct at the end of 2019.

The above efforts toward effective sand governance sought to implement a new cooperation at the national level to eliminate illegal sand mining, but illegal sand appropriation continued to take place despite these new national policies. According to T.-M. Li (2005, 391), "the effects of planned interventions have to be examined empirically in the various sites where they unfold." Yeh (2015, 623) furthered this argument by drawing attention to "how specific state interventions play out in their encounters with local agents whose subjectivities are shaped but not fully determined by state power." This book looks at such encounters on various levels in the case of illegal sand mining in southwest China.

2 The Current Demand for River Sand in China

China is arguably the fastest-growing economy in the world, with an annual growth rate of 6.8 percent in 2017 (NBS 2019). Economic growth is accompanied by processes of urbanization. In 2012, China's urbanization rate exceeded 50 percent (Wen 2012). This meant that China's urban population for the first time exceeded its rural population. In 2020, China's urbanization rate was 60 percent (Li 2020). Significant changes to all aspects of life have occurred during the current wave of urbanization. In terms of living standards, more than 6 million homes in rundown urban areas and over 3.8 million dilapidated rural houses were renovated (Keqiang Li 2017). The "Report on the Work of the Government" in 2017 details improvements in transportation and water conservancy:

> Over 1,900 kilometers of new high-speed rail lines came into service, and more than 6,700 kilometers of expressways and 290,000 kilometers of rural roads were built or upgraded. Construction picked up pace on urban rail transit facilities and underground utility tunnels. Construction began on 21 major water conservancy projects.
>
> KEQIANG LI 2017

Due to the overwhelming number of state-driven construction projects in both urban and rural areas, river sand is in high demand (e.g., Fidjett 2003; Jia et al. 2007; Wu 2008). According to a report on China's mineral resources by the Ministry of Land and Resources in 2012, the nation's mineral resources have made great contributions to the establishment of China's modern industrial system and provided support for the sustainable and healthy development of China's economy and society (MLRPRC 2012). Nonetheless, the disparity between supply and demand of natural resources is increasingly conspicuous (Fidjett 2003; Ho 2006; Haller 2010). Also, the demand for river sand has been rapidly increasing, and hundreds of millions of trucks have been rushing to transport river sand mainly from rural rivers to wholesale markets and various types of construction sites in rural and urban areas across China.

According to a 2014 report by the United Nations Environment Programme (UNEP), the demand for sand is greater in China than anywhere else in the world. The UNEP estimates the use of sand by analyzing the production of cement for concrete. One hundred and fifty countries produced 3.7 billion tons of cement in 2012. China contributes heavily to this development. The demand for cement by China has increased by 437.5% over two decades (1994 to 2012; see Figure 1), while its use in the rest of the world increased by 59.8%. Each Chinese citizen is currently using 6.6 times more cement than a U.S. citizen (USGS 2013a). For each ton of cement, the building industry needs about six to seven tons of sand and gravel (ibid.). This shows the relation between economic development (e.g., building booms) and resource consumption.

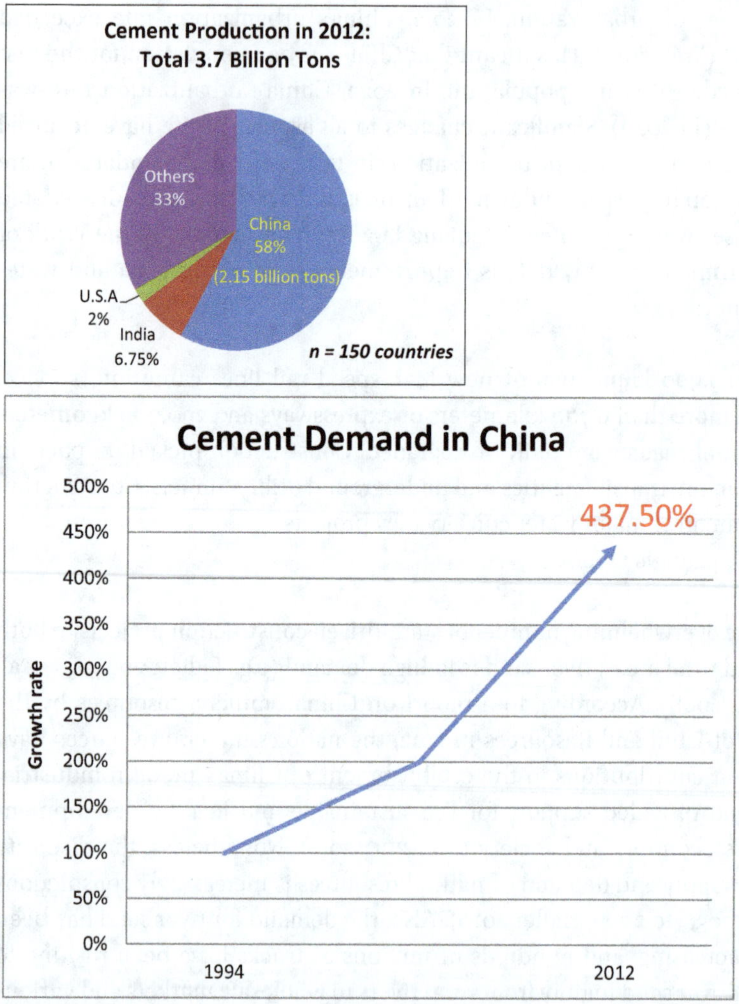

FIGURE 1 Data on the cement, sand, and gravel industry
SOURCE: USGS 2013A, 2013B, 2013C; GRAPHS CREATED BY THE AUTHOR

An article titled "The Secret of the Sand Shortage," published in China's *Southern Weekly* in 2019, indicated that

> 70,000 tons of sand and gravel are needed to build a highway per kilometer; and one kilometer of high-speed rail requires 80,000 tons of sand and gravel; about 800 kg of sand and gravel is needed per square meter in housing construction.
>
> SHIYANG LIU 2019

China's annual output of sand and gravel is 20 billion tons, making it the world's largest producer and consumer of sand and gravel in 2018 (MIIT et al. 2019). Sand and gravel have become a huge industry and are two of the most in-demand mineral resources in China. These resources are critical for China to achieve a better life for its people.

In the last three decades, China's industrial revolution, rapid urbanization, and program of "building a new socialist countryside" (*shehuizhuyi xinnongcun jianshe* 社会主义新农村建设) have all been mainly based on the use of local, cheap raw materials, cheap labor, and cheap imports from rural areas and other countries, including countries in Africa, Southeast Asia, and South America (Jiang 2009). Moreover, urban and rural growth have depended on China becoming an increasingly important foreign direct investment exporting country (Cai 1999). These developments inevitably have significant implications in terms of both national and international political ecology. In terms of sand as a resource, China had exported river sand to Japan, Taiwan, and South Korea for decades. However, current economic strategies and the rapid construction boom have resulted in sand resource overexploitation all over China. Moreover, due to the increasing constraints of environmental policy in recent years, for instance in the form of environmental protection and water resources protection policies, the ban on river-sand mining in the country has tightened (CSSGLW 2018a). The provinces and the authorities at and below the provincial level have carried out improvements of the River Chief System. Thus, sand mining in many rivers has been shut down. Since 2017, China has begun to import river sand from Southeast Asia (CSSGLW 2018b), and even from Africa (Mwesigye 2018), to supply the increasing domestic construction demand. For example, the Shanghai Association of Sand and Gravel has started to deal with high-quality sand and gravel suppliers and began importing river sand from the Pahang River in Malaysia in 2018. Meanwhile, according to the Report of the Chinese Sand and Gravel Association, the price of river sand in more than twenty areas has rapidly increased since the beginning of 2017; in some areas river-sand prices reached 220 yuan per ton in June 2018, compared to 110 yuan in June 2017 (CSSA et al. 2018). Indeed, Hunan Province's river-sand prices soared, increasing nearly sixfold, as the ex-factory price exceeded 180 yuan per ton (CSSA et al. 2018). The demand for sand in China has clearly increased at a fast pace over the past decades. Nevertheless, this phenomenon has hardly been studied.

3 The Political Ecology of River-Sand Mining

Political ecology has been a major approach for exploring natural resource management and common property, and its popularity has grown swiftly in the last three decades (Vayda 1983; Blaikie 1985; Blaikie and Brookfield 1987; Atkinson 1991; Hecht and Cockburn 1992; Robbins 2012). The intensifying environmental crises and resource impoverishment worldwide have stimulated scholars to seek a new lens to interpret the issue of environmental degradation. From 1970 onward, political ecology has played a crucial role in illustrating the complex causes of environmental degradation, resource exploitation, and political processes. However, given the worldwide crisis of land and water during the past several decades, political ecology has focused much on land, a bit on water, and less on other resources, including sand. This book contributes to the field of political ecology by adding a new object of research: river sand.

As I mentioned previously, there are several types of sand and numerous uses for it. River sand is especially interesting because it is somehow both part of land and water: it is connected to rock, land, water, and rivers and is part of the river ecological system. Moreover, it is also in great demand for numerous construction projects worldwide. Torres, Brandt et al. (2017) state: "Sand and gravel are the largest portion of these primary material inputs (79% or 28.6 gigatons per year in 2010) and are the most extracted group of materials worldwide, exceeding fossil fuels and biomass." This makes a focus on river sand a fitting lens for exploring ongoing human–environment relationships. The governance of river sand is both similar to and different from the governance of other resources. For instance, determining ownership and boundaries is easier for land and forests than for river sand. River sand is similar to water in that it is somewhat fluid, since it is carried by water, but it can also become part of the land. Thus, the governing of river sand must take political jurisdiction into account (see chapter 5). Moreover, river sand was a very rich and low-value natural resource until the construction boom in the twenty-first century. Thus, the formal institutions for governing river-sand mining are only now emerging, and new policy is still being made in this regard.

Political ecology consists of "empirical, research-based explorations to explain linkages in the conditions and change of social/environment systems, with explicit consideration of relations of power" (Robbins 2004, 12). It further explores "the relations between human society, viewed in its bio-cultural-political complexity, and a significantly humanized nature" (Greenberg and Park 1994, 1). Notably, political ecology has broadened its concern to cover a wide range of topics (Vaccaro et al. 2013, 255), including

> access to and control over resources; marginality; integration of scales of analysis; the effects of integration into international markets; the centrality of livelihood issues; ambiguities in property rights and the importance of informal claims to resource use and access; the importance of local histories, meanings, culture, and "micropolitics" in resource use.
>
> MCCARTHY 2002, 1283

Despite China's economic success, it has faced many severe environmental crises, such as water pollution, air pollution (Wong 2013), waste pollution, land degradation, biodiversity losses, cropland losses, depleted fisheries, seasonal flooding, drought, and health and social consequences (e.g., Shapiro 2001, 2012; Liu and Diamond 2005; Diamond 2011; Lora-Wainwright 2013; Stern 2013; Wong 2013; Lewis 2015). These environmental issues in China have severely impacted nature and the Chinese people. In the last decade, China has had to face the increasing scarcity of its river sand and the negative sociocultural and ecological consequences of river-sand mining.

Ho (2006, 3) points out that "China's development poses the greatest environmental challenge for the modern world in terms of speed, size, and scarcity." He cites three factors in particular:

> China's large population; its explosive economic growth and the increasing socio-economic cleavages which result; and its relative shortage of natural and energy resources. These three factors together imply that China will have a profound environmental impact at the global level. (11)

Nonetheless, the severe environmental consequences of river-sand mining are unequally distributed all over China. While river sand has been overextracted in the Yangtze River and Huai River in eastern China, some other rivers have just started to experience illegal sand mining, such as the Left River, on the border between the GZAR and Vietnam. Thus, river-sand mining has various influences in different places depending on geographic location, economy, culture, ethnic composition, religious beliefs, and the local sociopolitical situation.

Importantly, political ecology has also explored conflicts and social movements, institutions, capitalism, identity, entitlements, gender, risks, vulnerability, and resilience. These themes, of course, also connect to "a range of actors, typically including agrarian communities, indigenous peoples, state agencies, non-governmental organizations (NGOs), corporations, and global and transnational governance institutions" (Neumann 2015, 391). This complexity of themes and various actors, as well as their different places, has been studied in a broad range of empirical contexts.

Indeed, political ecology provides "key tools" and an important alternative framework to scrutinize sand issues by illustrating "the contention, power, institutions and scales" that are involved (Walter 2014, 21). Yet this approach has rarely been used in China's context, where environmental woes and resource scarcity are burdening development within the nation and can have severe effects beyond China itself (Ho 2006). Therefore, China's situation increases the demand for critical research in the field of political ecology to explore the complex relations between economic development, rising environmental upheaval, and environmental governance (Xie 2015; Yeh 2015). Nonetheless, "among the dozen major edited volumes and textbooks on political ecology published in English since the mid-1990s, only two include case studies from China" (Yeh 2015, 619). My own literature review shows that more and more case studies in the field of political ecology are set in China. In addition to works in English by Hershkovitz (1993), Tilt (2007), and Yeh (2000), for example, there are many works written in Chinese.[3]

Yeh (2015, 619) further points out that "such limited attention is disproportionate to China's weight in the world, whether geopolitically, economically or in terms of its territory and population." Yeh argues:

> This lacuna is undoubtedly related to the contingencies of scholarly lineages and networks. In the USA, China geography as a subfield emphasizes urban and economic geography, rather than political ecology or political geography. But the lacuna also reflects Chinese state control on scholarship, which affects foreign researchers' access while encouraging scholars in China to use the language and analytical lenses of the state. (619)

This control will hinder scholars' ability to figure out the numerous reasons for the vicious cycle of environmental crisis (Economy 2007; Smil 1997) and the impoverishment of the major resources in China (Chen and Xu 2017). In the end, more detailed ethnographic work is needed (Xie 2015; Yeh 2015). This book specifically pays attention to the rural spaces and to the role of local governments and communities. River sand plays a crucial role in riparian communities (Clark 1996). Rivers and river sand have long been considered a communal property of riverfront Zhuang villages (Zhu 2018, 2019). The local riverbank dwellers link their sociocultural life to rivers (Liu 1995; Marks 1996,

3 When I typed "political ecology" (政治生态学) into Baidu Scholar (the Chinese equivalent of Google Scholar), I received 465,000 results, including Duan et al. (2000), Liu (2007), Barry and Yang (2009), and Zhang (2015).

2004). From the historical perspective, rivers and river environments have been shaped by these villages' activities. Meanwhile, river landscapes have shaped the riverine society and culture as well. Under these circumstances, river-sand mining has caused many disputes and conflicts in those rural Zhuang villages where river-sand extraction occurs.

In terms of river-sand mining management and its consequences in China, Yuan and Liu (2000), Li (2006), Wu (2008), and Hou (2011) studied the river-sand crisis by focusing on the weakness of sand-mining institutions and management in China. Jia et al. (2007, 197–198) studied the impacts of the large amount of sand mining on riverbed morphology and tidal dynamics on the lower reaches and delta of the Dong River in Guangdong Province and concluded that sand dredging caused a substantial impact on riverbed evolution, tidal level, tidal wave spread, and other hydrological changes of the river. Other impacts of sand mining have also been pointed out with regard to various rivers in China (Han et al. 2005; Mao and Huang 2004; Zhang et al. 1996). Nonetheless, these writings either focus on environmental (e.g., physical, chemical, and biological) impacts, or on political and social consequences. A political ecology approach can combine both.

By applying a political ecology approach, this book further explores the dynamics of ownership regarding access to and control of river sand, with specific reference to Qinzhou City in the GZAR. Formal institutions (laws and government) have often been blamed for ineffectively managing the river sand. The efficacy of these institutions in controlling illegal operations and in mediating disputes is questionable (Wu 2008; Hou 2011; H. Li 2006). This book will also demonstrate the interconnected effects of current economic changes, cultural dynamics, and politics on natural resource management.

The Chinese constitution is ambiguous about the details of the ownership, protection, and management of river sand. According to the 1954 Constitution (Article 6), and in the 1982 revision under the Deng Xiaoping administration (Article 9), all natural resources fall under state ownership, which by extension was described as therefore under the ownership of the people. "The state ensures the rational use of natural resources and protects rare animals and plants. The appropriation of or damage to natural resources by any organization or individual by whatever means is prohibited" (CPRC82 1982, Article 9). Based on the Resources Law of the People's Republic of China, natural quartz sand belongs to the non-metallic minerals, which are owned by the state. Article 10 of this law also states:

> In mining mineral resources in national autonomous areas, the State should consider the interests of those areas and make arrangements

> favorable to the areas' economic development and to the production and well-being of the local minority nationalities. Self-government organs in national autonomous areas shall, in accordance with legal provisions and unified national plans, have the priority to develop and utilize in a rational manner the mineral resources that may be developed by the local authorities.

The issuing of sand-mining licenses is the major approach in governing state-owned river sand. According to Article 39 of the Water Law, "measures for implementing the licensing system for sand quarrying in river courses shall be formulated by the State Council." Moreover, sand-mining licenses can be issued by any level of government at or above county level, and there is great flexibility in the ability to grant sand-mining licenses. In this book, I understand sand miners with a governmental license to be legal miners, while unlicensed miners are considered illegal.

The volume of sand being extracted is having severe, multiple impacts on rivers, lakes, deltas, coastal, and marine ecosystems worldwide (Galay 1983; Kondolf 1997; Galero et al. 1998; Norman et al. 1998; Kondolf et al. 2002; Padmalal et al. 2008; Sreebha 2008). According to UNEP (2014), sand and gravel are now being extracted at a rate far greater than that of their renewal. This book describes how the continuous in-stream illegal sand mining in southwest China, and in particular the overexploitation of riverbeds and riverbanks, has changed the physical structures, waterways, and appearance of rivers and has resulted in damage to river ecosystems, the loss of fish species, loss of land, water pollution, and riverbank collapse. Despite these severe consequences, little attention has been paid to river-sand mining and its consequences worldwide (UNEP 2014; Sreebha 2008).

In rural China, the situation regarding individual usage rights to river sand is different. It is important to note that the central government decentralized the power to manage river sand by allowing the regional government to establish regional laws and institutions for sand management. The GZAR established its sand-government institution in 1992. According to the reform record, the sand-government institution has been renewed every ten years to take into account the changing circumstances of sand mining. In November 2016, the Autonomous Regional People's Congress adopted the Guangxi Zhuang Autonomous Region River Sand Mining Management Regulations. These were implemented on January 1, 2017. The regulations say that rural residents can take no more than 100 cubic meters of sand for self-built houses, and that to satisfy the demand for the construction of public infrastructure, a village must rely upon the river for sand mining, which is to be conducted outside restricted areas and which is not subject to either mineral-sand licenses or sand-mining fees (SCPCGZAR 2016, Article 14).

The right to use river sand is considered a crucial right of rural residents and rural villages in the GZAR. The long-term customary right to use sand cannot be eliminated by law-governing institutions in Guangxi. This is because many rural Zhuang villages are located in close proximity to rivers, and many rivers are part of the village communities' territories. Rivers and river sand have long been considered as a part of territory and a communal property of riverfront communities. Local riverbank dwellers have long interlinked their sociocultural life with rivers within their territories. Thus, many residents of the Zhuang villages claim that "this is our sand." I found that there are many cases of local resistance against illegal river-sand mining, including resistance through violence, resistance through rural formal government channels, resistance through petitioning the government, and resistance through social media.

This book examines how different stakeholders are involved in river-sand extraction. It also explains different types of user rights and discourses in the Zhuang villages of southwest China in terms of uncontrolled resources. Furthermore, it investigates what kinds of resource-management institutions have emerged to deal with the control of natural resources, how they emerged, and why these institutional dynamics of river-sand governance in Southwestern China have resulted in a fragmentation of power.

A few scholars have studied sand mining and the river system. G. M. Kondolf is an outstanding example. Kondolf's writings (1993, 1994a, 1994b, 1995a, 1995b, 1997, 1998a, 1998b) emphasize the issues of sand mining, riparian zone management, and stream channel processes as they have related to natural resource management in California and beyond since 1993. Dissertations have also been written on sand-mining issues. For instance, S. Sreebha's dissertation, "The Environmental Impact of Sand Mining: A Case Study in the River Catchments of Vembanad Lake, Southwest India" (2008), explains the linkage between sand mining and the ecological degradation of rivers and lakes due to population growth and economic development. A. Krishnakumar's (2002) research focuses on the environmental degradation of two river basins in southern Kerala in India.

In this book, I apply the framework of political ecology and consider power relations in the management of the natural resource of river sand. Political ecology provides a crucial lens and key tools to explore river-sand mining issues by illustrating the multiple actors involved and by investigating conflicts, river degradation, indigenous knowledge, and human-river interaction. I highlight some of the power dynamics between different stakeholders and institutions at various scales, as well as the disputes arising between them. In so doing, I identify some of the winners and losers of illegal sand mining. I distinguish between four main types of actors, who all have a different perspective on ownership and user rights of river sand:

A. The state, which is the legal owner of the sand.
B. Zhuang villages, which claim ownership of or user rights to the sand.
C. Illegal sand miners, who claim the sand belongs to nobody.
D. Individual households, which have a legal right to use no more than 100 cubic meters of sand for free according to laws in Guangxi.

Significantly, only a few ethnographic studies have applied a political ecology approach in the context of China. My book expands the existing literature by applying political ecology to the crisis of river-sand mining in southwest China.

My intention is to contribute a detailed ethnographic work to further the academic debate. Combining politics and ecology (Bryant and Bailey 1997; Robbins 2012), rather than taking an apolitical (Forsyth 2008) or depoliticized approach (Swyngedouw 2015), provides a new lens to look at the critical problems of the river-sand crisis. This study will address the interactions between various actors, political institutions (Peet and Watts 1996), narratives and stories (Escobar 1996), and environmental change (Watts 1985), yielding new insights into human–environment relationships in modern China.

4 The Concept of Indigeneity in China

In this era of global environmental crisis, indigenous ecological knowledge (e.g., traditional philosophy/religion/wisdom) has attracted considerable interest as a potential means to deal with specific environmental issues (Aswani, Lemahieu, and Sauer 2018; Duara 2015; Von Glasenapp and Thornton 2011; Senanayake 2006). Indigenous ecological knowledge is generated and owned by indigenous groups. There is neither an official definition of "indigenous" adopted by UN-system bodies (UN 2015), nor is there a clear consensus on the meaning of the term among scholars (Jackson and Warren 2005). In his field research on a Tibetan village in Qinghai province, Yü (2017, 125) found that "the meaning of being indigenous has evolved with changing historical and religious circumstances and conditions." Hathaway (2010, 303) believes that "understanding indigeneity in any social context can be furthered in part by regarding indigenous people not as a natural identity but as a social and political category." Furthermore, this political category of indigenous people is influenced by transnational connections and global developments.

Nonetheless, the term "indigenous" has been imported and then applied in the context of China mainly by three paths. First, whereas the Chinese government considered indigeneity an irrelevant concept in China, some international environmental NGOs, such as the Nature Conservancy and the Worldwide Fund, "pave[d] the way to discover and document an indigenous presence" in China since the 1980s (Hathaway 2016, 13). Second, as argued by Hathaway (2010), Chinese intellectuals have invoked the term "indigenous people" in China in the service of diverse causes since the 1980s, such as in a scholarly debate on the importance of sacred lands in maintaining forests and promoting biodiversity in Yunnan Province (Pei 1985, 1993) and when discussing the role of indigenous women in the process of natural resource management (X. Yu 1993). They challenged the stereotypical idea that slash-and-burn practices by upland people led to deforestation in China and built a more serious classification of indigenous knowledge in agricultural practice (Xu, Pei, and Chen 1997; Xu, Fox, et al. 1999). Third, Chinese overseas students and international scholars also applied the term to discuss the relations between communal resource management, climate change, conservation, and indigenous knowledge in China (Yeh 2007; Sturgeon 2007; Yin 2015; Q. Zhu 2020).

Through these three ways, the vital role of the rich indigenous knowledge of China in sustainable development has attained increasing recognition, in line with global trends (Senanayake 2006; Aswani, Lemahieu, and Sauer 2018). The indigenous knowledge of China's ethnic minorities contributes to strengthening nature conservation and promoting biodiversity (Yin and Zhang 2020), while global trends show loss and change of local and indigenous knowledge "driven by globalization, modernization, and market integration" (Aswani, Lemahieu, and Sauer 2018, 1). This loss of traditional ecological knowledge has also been documented for indigenous communities in the Chinese province of Yunnan. Yin et al. (2020, 63) state that

> policy-led large-scale development of the rubber industry in Xishuangbanna not only causes the quick disappearance of biodiversity, but also leads to the break off of the traditional ecological knowledge of ethnic minorities, and finally causes simplification and disappearance of cultural diversity.

The Chinese government denied that the country has "indigenous peoples" in the sense of colonized people. Nevertheless, given the growing importance of indigenous knowledge in global biodiversity conservation and environmental sustainability in the last several decades, indigenous knowledge is used as a theoretical framework and sometimes also as a policy-making guideline to study the traditional wisdom of Chinese ethnic groups and the crucial contributions of their indigenous knowledge to biodiversity conservation and sustainable management of resources (Xu, Ma, et al. 2005; Jiao, et al. 2012). In the face of increasing attention to indigenous interests and rights, China argues that its ethnic groups, including the majority Han, are indigenous (see also Hathaway 2010, 302). Hathaway's (2010, 2016) research in Yunnan Province showed there are neither grassroots movements of indigenous peoples, nor do people identify themselves as indigenous in Yunnan. Even though they as Tibetans "most easily attain recognition as indigenous by international organizations, … they are either largely unaware of the possibility or uninterested in this status" (Hathaway 2016, 1). Furthermore, "rural citizens—including those who would most likely be deemed indigenous by outsiders—have never heard of the concept" (Hathaway 2010, 322). Thus, the term is rarely used by rural groups themselves in China (Sturgeon 2007).

Actually, not only has the meaning of the term "indigenous" shifted in various places and countries, but the Chinese term *minzu* (民族)[4] has also had various meanings, including nation, ethnicity, nationality, and people. Its primary English translation is the term "nation," a concept that was originally created in the West in the eighteenth century. The idea of the nation was imported to China in the early twentieth century (Bilik 1995, 9). Even though there was no clear consensus on the exact meaning of the term, China's *minzu* theory was mainly impacted by Stalin (Wang 2015, 7). Based on Stalin's four elements of a nation,[5] China conducted its Ethnic Identification Project in the 1950s. The project proclaimed that the Chinese nation officially comprised fifty-six nations who were politically equal according to the national constitution. By 2010, however, the national census showed that there were as-yet unidentified groups with a total population of 640,101 (Wang 2015, 16). Yang (1986) argued that Stalin's "nation" concept did not suit the *minzu* situation in China. According to L. Wang (2015), the application of the concept of nation or nationality was ended after the dissolution of several former socialist states in the 1990s. "The State Nationality Affairs Commission became the State Ethnic Affairs Commission to avoid any possible political implications deriving from the concept of nationality" (3–4). It is interesting to learn that there is no consensus on the definition of *minzu* among Chinese scholars (Yang 1986; Bilik

4 People mostly use the term *minzu* in formal situations, such as school enrollment, university admission, household registration, and marriage registration. See also L. Wang (2015).

5 The four elements of a nation are "a historically constituted, stable community of people, formed on the basis of a common language, territory, economic life, and psychological make-up manifested in a common culture" (L. Wang 2015, 7).

1990, 1995; Wang 2015). It seems that most Chinese scholars are still searching for a better English translation for the term *minzu*, while others have translated it as "indigenous" or "indigenous groups" in diverse contexts (e.g., Yü 2017 or Yin 2015).

At a time of global environmental crisis, the application of the concept of indigeneity to an understudied place and an understudied ethnic group in southwest China contributes to creating a new academic discussion in the Chinese context and adds an updated empirical case study to the literature of global indigeneity. This may help decision-makers to establish new laws and policies with the aim of enacting social change. Thus, this book applies the term "indigenous" to describe the Zhuang people, who neither positioned themselves as indigenous nor identified themselves as Zhuang people during my fieldwork. They mostly referred to themselves as "local people" and simply used "us" to present themselves during the struggle processes. I follow the idea that "the category of indigenous people exists as a social fact … a concept that fosters particular social worlds" (Hathaway 2010, 303–304). Some of these social worlds can in turn mobilize people to look for alternative futures (Hathaway 2010, 322). Therefore, I apply the term "indigenous" to the social world of rural Zhuang villagers and their knowledge.

Following Yin's (2015) discussion, within the Chinese context, indigeneity transforms into a concept that is interchangeable with ethnic minority, and I therefore use the terms interchangeably when describing my specific case. This book explores the interaction between the indigenous knowledge of the rural Zhuang communities and natural resource extraction by consistently applying the concept "indigenous" to the largest Chinese ethnic minority, which has dealt with severe environmental problems and public resource management dilemmas.

5 The Zhuang and Their Indigenous Ecological Knowledge

As in many developing countries, economic growth in China is accompanied by unsustainable development and environmental crisis. The question is: Can the ecological knowledge of minorities help to construct a sustainable future in an era of booming construction and a river-sand crisis? Some scholars believe that indigenous knowledge may contribute to solving the crisis of natural resource management and environmental degradation in rural minority areas. For example, Duara (2015, 2) argues that the model of Western modernity, which is deeply rooted in Max Weber's assessment of Protestantism, has resulted in "the crisis of sustainability." Duara then aims to "revisit the alternative traditions from China and India, many strains of which have adapted to the unceasing circulations of modernity, to examine whether they allow a more viable cosmological foundation for sustainability" (2). Duara (2015, 280) argues:

> Local ideas of sacred place and nature continue to abound across Asia and many other parts of the global south; and while they are susceptible to denial and abuse, as we have noted, they remain an important source for regenerating nature. Indigenous and local knowledge are in dialog with scientific techniques of agroforestry and disaster recovery; cultural and biological diversity are being found to be intimately connected.

Moreover, the role of non-Western philosophy/religion has been increasingly highlighted. Anderson (1996, 166) observes, "All traditional societies that have succeeded in managing resources well, over time, have done it in part through a religious or ritual representation of resource management." The linkages between major religions/philosophies/cultures and their special worldview and value of nature have been explored worldwide (Hargrove 1986; Rockefeller and Elder 1992; Callicott 1994; Kinsley 1995). For instance, Rappaport (1967) wrote about "Ritual Regulations of Environmental Relations among a New Guinea People." The ritual is the regulation that "helps to maintain the biotic communities existing within their territories, redistributes land among people and people over land, and limits the frequency of fighting" (17). The ritual is the regulator that monitors the "relationships between people, pigs, and gardens," and the "slaughter, distribution, and consumption of pigs," balancing the quantity of people and nature in the territory (Rappaport 1968, 3–4). In "The Historical Roots of Our Ecological Crisis" , White (1967) states:

> What people do about their ecology depends on what they think about themselves in relation to things around them. Human ecology is deeply conditioned by beliefs about our nature and destiny—that is, by religion.

The increasing awareness of the importance of indigenous knowledge in environmental conservation is mainly a criticism of Western philosophy. Paper (2001, 6) mentions the Western philosophical and religious traditions as the basis of our present plight. However, is original ecological knowledge able to solve the situation? Existing studies (Huang 2003a, 2003b, 2003c; Zeng 2008; Fu 2014) have argued that the ecological knowledge of the Zhuang is profound and may contribute to the solution of ecological problems.

I try to link sand exploitation near Zhuang villages in rural China to indigenous ecological knowledge by illustrating how local knowledge applies to river-sand mining in the context of southwest rural China. Furthermore, the chosen six villages along the Maoling River belong administratively to two Zhuang-dominated towns (Huangwutun Town and Xintang Town) administrated by Qinzhou City in the GZAR, where Zhuang history, culture, resource management, and ecological knowledge both shape and are shaped by the river; thus, the indigenous ecological knowledge of the Zhuang is rather special in terms of river-sand mining in their territories.

The Zhuang ethnic group is China's largest minority group. Nearly 86 percent of the Zhuang population live in the same region in Southwest China, therefore the Zhuang were given their own autonomous region on March 5, 1958, by the Chinese central council based on "nationalities work" (GBSYC 2011). This region formed one province, the Guangxi Zhuang Autonomous Region, generally known as Guangxi. Even though the Zhuang nationality is referred to as the largest ethnic group in China among the fifty-five minority groups, there is little academic research on the Zhuang. According to Kaup (2000, 25),

> The Zhuang nationality is almost entirely unknown to Westerners. Few non-Sinologists have even heard of the group, and those scholars who mention the Zhuang at all in their analyses overwhelmingly dismiss them as "fully assimilated" and essentially no different from the Han majority. … [The Zhuang] as a minority group actually do not pose any minority problem for the Chinese administration.

In this respect Guangxi stands in contrast with the best-known autonomous regions, Tibet and Xinjiang. Kaup (2000, 1–2) notes that Guangxi is a rather peaceful and prosperous agricultural region and has been a good example of cooperation with, or obedience to, the Chinese minority policy, but "Tibet and Xinjiang Provinces in particular have been intermittently rocked by violence since the Communists first declared them 'liberated' shortly after ousting the Nationalist forces in 1949" (2).

There are only a few Western scholars (Kaup 2000; Lary 1972; Took 2005) who have focused exclusively on the Zhuang. Furthermore, the indigenous ecological knowledge of the Zhuang was ignored by scholars for many years, until signs of crisis in the ecological system began to emerge. Some Chinese scholars, in particular Zhuang scholars, have shifted their attention to the value of the Zhuang's ecological knowledge in constructing an "ecological community" and an "ecological Guangxi," as argued in Qin's (1999) paper. Other aspects of the ecological environment of Zhuang culture, including *Na* culture (*Na* 那 means "paddy field" and "land" in the Zhuang language), have been studied by N. Qin (1997, 68–69), C. Qin (1998, 40–43), Deng (2012), Ling (2010), and

Zeng (2008). *Na* culture could be understood as "rice culture" (*daozuo wenhua* 稻作文化) or "agricultural/land culture" (*nongye huo tudi wenhua* 农业或土地文化). In Guangxi, many names of towns and villages begin with *Na*, such as Nagan and Naxia—two villages in this study's sample (Qin 1997). However, works on Zhuang ecological knowledge have not explored, for instance, its connection to settlement patterns, house styles, or the current environmental crisis in the region.

The Zhuang residents of Qinzhou have a tradition of living in a lineage-based group, but they differ greatly from the Han people regarding the choice of location and architectural style of their settlements (Li and Pang 2000; QLCCC 2018). I argue that the new style of houses introduced into rural Guangxi demanded river sand, cement, and other materials, a demand that has contributed to the boom in river-sand mining.

The link between lineage organization and settlement (land) is a key feature in south China (Freedman 2007; Liu 1995; Siu 2016). Freedman (2007, 5–6) pointed out that "there is a temptation to look upon the single-lineage settlement as the historically prior form and mixed settlements as evidence of a later disturbance of the primordial pattern." Kaup (2000, 45) stated about Guangxi: "Generally speaking, the Zhuang and Han live in the most fertile valley areas and grow rice as the main crop. The Yao, Miao, and Yi, which are the groups in closest proximity to the Zhuang, tend to live in the more mountainous areas." An old saying about the locations of each ethnic group goes: "The Miao and the Yao live at the head of the mountain; the Zhuang and the Dong live at the head of the river, and the Han stay at the head of the street." The Zhuang villages of Qinbei District are mainly located along the Maoling River and its tributaries, where they are situated beside the mountains or hills, close to the fields and near the river. There are similar single-lineage settlements among Han people and other minority groups in Guangxi. Before liberation (December 1949), Zhuang residents used stones to form a wall around the village, planted thorns outside the wall, and built a gatehouse at the entrance of the village, setting up fences in the gate building; some villages also built a fort, standing in the middle or at the side of the village, to facilitate observation and defense (Li and Pang 2000).

Decisions on where Zhuang villages are located are affected by various issues, such as the number of residents, natural resources, religious practices, ethnic identities, agricultural technology, and the level of education and business skills of the villagers, as well as the history behind these issues. Furthermore, the distribution density of the villages also depends on the capacity of the environment. That is, it follows the basic principle of having sufficient land for subsistence. The villages are generally about one kilometer apart. Around each village, people often plant a variety of bamboo trees to create a green environment to prevent soil erosion (Li and Pang 2000). As discussed below, the Zhuang villages have at least one big tree in their territories where their collective worship activities take place.

Housing in the Zhuang villages on which this study focuses has undergone a major transformation in the past thirty years, and this transformation is closely related to river-sand mining. Housing styles in Zhuang villages have become modernized, reflecting the modernization of households and the modern lifestyle of peasants in the villages. Yet, the cement-and-steel houses with big glass windows in a modern style demand much more sand than wooden-beamed traditional houses. This is reflected in the increase in the value of river sand and other construction materials in rural areas. River sand as a material has been rapidly depleted by utilizing new mining technology in rural areas. The architectural transformations in the Zhuang villages reveal the influence of state-driven rural modernization. Meanwhile, the peasants' lives and their capacity to build new homes have been impacted by their religious and sociocultural customs and natural resources. The drive to build new homes has altered the balance of the demand and supply for river sand in Zhuang communities. As a result, river-sand mining activities have played a major role in the political-cultural and economic-environmental transformations in these rural areas. This interaction between rural villagers and river sand illustrates the mechanism of ecological resource degradation in the construction era in contemporary southwest China. Furthermore, it is crucial to explore the role of the belief system of the Zhuang ethnic group in the crisis of river-sand mining, since a rural Zhuang community is a faith-based and lineage-rooted network system. Rituals bring people together and inspire villagers to take action against illegal sand mining.

The Zhuang created numerous works of ancient literature, and I highlight the creation epic *Buluotuo* (布洛陀) here, which illustrates Zhuang cosmology. "Buluotuo" is a transliteration of a Zhuang term meaning "the head of the mountain," "the old man in the mountains," or "intelligent old people who know everything." The epic is full of enthusiastic praise of the Zhuang ancestor Buluotuo, who is half-human and the ancestor of the demigods (NEAC 2018). According to the *Buluotuo*, the thunder king (the heavens), the dragon king (the earth), and the first ancestor of the Zhuang were born from the same stone. The epic points out that the heavens, the earth, and humans are like brothers who share the same mother. This "mother" was a stone that condensed from black, yellow, and white gases. Therefore, gods and humans are from the same source (Ling 2010; Ouyang 1986). Ling (2010, 90) sums up the message of the epic thus: "To seek a harmonious and orderly development of human beings and nature, people are required to comply with nature, respect natural rules, treat nature nicely, respect life, and establish a friendly relationship between human beings and nature."

Since ancient times, the Zhuang have been polytheists, believing in the power of many inanimate things in nature, such as giant trees, high mountains, the earth, the sun, stones, rivers, frogs, and their ancestors (Ling 2010, 91; Deng 2012). The Zhuang regularly engage in activities that include making offerings because they believe that when they do so they are blessed by the divinity, thus preventing all kinds of disasters. For instance, the Zhuang maintain a large tree around a temple on their land where land gods are worshipped. This worship symbolizes the prosperity of the village. Trees are also worshipped during the festivals, and are considered the spiritual fathers or mothers of some of the villagers (Xu 2006; Wang 2003). It is forbidden to cut down these trees or even to urinate around them, and it is believed that violators will be punished by the gods. This belief system has been maintained for generations, and the Zhuang gather for these religious activities (Liao 2004). The Zhuang realize that the sacred trees, forests, and frogs have a positive function with regard to their farming, as they influence the rainfall and environment (Ling 2010).

The spiritualized nature of the Zhuang's ecological knowledge is similar to that of other indigenous groups, such as the eco-religious knowledge of Tibetans, as clearly shown by Yü (2014, 2017), Yin (2015), and Yin et al. (2020). These scholars find that the presence of the sacred and the deity mountains form the center of the Tibetan religious system, entwining landscape, human beings, and gods. Furthermore, many discourses have been built upon this idea in order to protect the mountains as well as the trees and animals on the mountains. Yü (2014, 2017) used two terms to describe the Tibetans' sacred religious beliefs (or practices), namely "an emplaced religion" and "an indigenous cosmovision," both referring to the special relationship between religion and ecology/environment in the Tibetan worldview. Yü (2017, 123) highlights that the Tibetan's sacred sites are seen "as multidirectional connectivities with gods, the earth, humans, and other sentient beings, not just gods alone. They signify an intersentient communication." Yin et al. (2020, 57) find that many other ethnic minorities in Xishuangbannna in Yunnan Province also "view some mountains as Sacred Mountain(s). Trees on these Sacred Mountains have been protected well by people until today." The concept of "sacred" environmental elements is also crucial in the Zhuang people's belief system and has an impact on people's lives and worldview.

In the chosen research settings, Huangwutun Town and Xintang Town, ancestral worship, land-god worship, ghost worship, and the geomancy system (also known as Fengshui, literally the "wind and water system") constitute the major belief system. The ancestors are considered the gods of the household, who protect and provide benefits for their offspring if the ancestors are buried in a good geomantic position in the mountains and if the house is located in an appropriate environment according to the rules of Fengshui. The land god is considered the god of the village, who protects the harvests and the people of the village. Ghost worship is another side of worship, practiced when bad things happen. The villagers then visit a wizard to deal with the problems that ghosts have caused.

The villagers collectively worship the god of land four times each season, while households may also worship at the temple of the god of land on any festival and during any ceremonial activities such as weddings or funerals. These kinds of worship are so normal in everyday life that they demonstrate the continual interaction among the gods, the village, its households, and the individuals in the village. Indeed, ancestors and the land gods play a crucial role in the villagers' daily lives. Each household carries out a prayer ritual at the beginning and the end of the year, during which rice is added to a specific container (*tianliang* 添粮, which means adding a longer life or a blessing for a person) for the elderly, mainly for parents and grandparents. The Zhuang believe that when an elderly person reaches a certain age, his or her "food" (meaning life energy) may be "used up" (which would result in their death). It is necessary then for his or her children, grandchildren, and married daughters- and sons-in-law or other close relatives to make offerings of rice, grain, or money (*tianfu tianshou* 添福添寿). Rice is frequently used in most of the rituals conducted by the religious masters in the Zhuang region (Chen 2011).

In my own experience, household worship during the New Year is also crucial in claiming property. When I was a child, after the domestic worship I had to stick red paper to each litchi tree, pig ring, and toilet, and to a bullpen that belonged to our household. People believe that when this practice is observed, their gods and ancestors will protect their belongings and bless the households' harvest and income. Nowadays, this red paper is also stuck on new cars and boats for the same purpose.

This belief system has been maintained for generations. I observed that collective worship helps the local people in many ways. First, the worship system maintains common activities by gathering most of the villagers together, even those who have migrated away from the villages. The religious activities are still the duties of each household in each village. Second, this system also serves as a way of dealing with all the conflicts and common issues regarding the village's collective situation. Third, it is a platform to discuss the collective affairs of the village. From an ethnographic point of view, paying attention to the belief system contributes to a better understanding of resource conflict and local resistance in China.

Festivals are integrated into the Zhuang people's beliefs, customs, food culture, and entertainment, all of which play a key role in people's personal and collective lives. There are various kinds of festivals based on either the village unit, lineage unit, or household unit. Sometimes there is no clear boundary between village-based and household-based festivals because they overlap and interact.

Nowadays, the GZAR has designated March 3 (lunar calendar, *sanyuesan jie* 三月三节) as the start of a three-day provincial festival holiday. March 3 is not only the birthday of the Zhuang ancestor Buluotuo, but has also become the day of a regional and traditional singing festival. The famous village-based and household-based festivals are known as "dancing on the head/slope of the mountain" (*tiao lingtou*, 跳岭头), "worshipping the common ancestral temples and graves" (*jizu* 祭祖), and "worshipping the Land of Gods" (*bai tudi* 拜土地). The Tiao Lingtou, also known as the Lingtou festival (*lingtou jie* 岭头节), is the most famous festival in Qinzhou City, and it mainly takes place at the level of the natural village.[6] The major activity is dancing by the masters (*tiao* means "dancing" in the local language), accompanied by local drummers. This is a traditional folk festival of the Zhuang and the Han celebrated during the Mid-autumn Festival in August (*zhongqiujie* 中秋节) in the Qinzhou area.[7]

---

6 When I talk about "villages," I refer to natural villages (*zirancun* 自然村) rather than administrative villages (*xingzhencun* 行政村). An administrative village consists of several natural villages, ranging from six to ten in the region of my fieldwork. Thus, a natural village is a subordinate unit of the administrative village and includes one to three production teams (*shengchandui* 生产队). "Administrative unit" is the name given to the lowest-level unit in the governmental hierarchy (Pei 2005, 209). Production teams were established in the late 1950s; they are an organizational form in China's socialist agricultural economy, consisting of about ten households. A production team is a means to gather people to work effectively and was also a work unit in the Mao era, when households were divided into diverse teams based on population. As Ho (2005, 209) explains, "A three-level organization was devised for the commune system: collective ownership rights were divided among the commune (the township since 1984), the brigade (the (administrative) village since 1984), and the production team (natural village). The team, consisting on average of 20 to 30 households and farming 15 to 20 hectares of land, was the primary unit of ownership, production, distribution, and accounting from the early 1960s to the early 1980s. The team members worked jointly on the land, and their remuneration was solely based on the labor they contributed to team production. This team system was replaced by the household responsibility system (HRS) in the early 1980s. Under the HRS, land is contracted to individual households and the rights to use it are separated from collective landownership, which is retained by the village."

7 The "Tiao Lingtou" festival was selected for the fourth batch of national intangible cultural heritage representative projects in 2014, marking a new step in the protection of the

The official records of Qinzhou show that the festival has been celebrated for 470 years (Li and Pang 2000; Lin 2007).

When I was a child, I participated in the Lingtou festival in Dawan Village in Xintang Town, where my aunt lived. The villagers explained: "Before the Tiao Lingtou celebration, the villagers will be fishing in the village ponds, and the households will sacrifice ducks, chicken, or fish to the ancestors and the gods." This is the same in Huangwutun Town. Such collective actions need to be well-organized under the control of traditional institutions, such as fund-collecting and voluntary labor organizations. On the one hand, conflicts and disputes occur among the members. On the other hand, the collective banquets and drinking together help them to build a better-integrated neighborhood and brotherhood. The Lingtou Team (religious masters) meet at the mountain slope of each village to carry out dance activities and spiritual cleaning rituals, and to pray for peace and a good harvest. The masters, who are local peasants like the other villagers, wear masks and perform *tiao* and also sing songs and enact dramas. There are also martial arts and folk magic performances. The performance program includes setting up altars, inviting the gods, and exorcisms (catching evil spirits and sending them away in a dragon boat, burning the boat as it floats along the river). There is a drumming accompaniment to the singing and dancing (QLCCC 2018; Lin 2007).

In addition to the Tiao Lingtou festival, the religious event "Worshipping the God of Land" (*bai tudi* 拜土地), or Zuo She (做社), followed by "Eating Tudi" (*chi tudi* 吃土地), which means having a banquet together in the temple of the land gods, is also very popular. Furthermore, the Qingming festival (清明节) and Chongyang festival (重阳节) are important lineage- and village-based ancestor-worshipping activities in the areas studied in this book. Other regular festivals include those for ghost worshipping, such as the ghost festival (zhongyuan festival 中元节). During these events, participants pray for a blessing for the agricultural harvest, more job opportunities, safety, and smooth progress for a better life.

Collective events are integrated into household activities with respect to funds, labor, participants, and performances. The festivals thereby answer many needs of (and require much engagement by) both collectives and individuals in the villages examined here. Yet the festivals also put pressure on the villagers to find more funds, for instance through sand-mining or other activities that exhaust the natural environment.

---

intangible cultural heritage of the city of Qinzhou. It has now evolved from a religious festival into a form of folk entertainment in some tourist destinations (QLCCC 2018).

6 The Objective, Research Questions, and Structure of This Book

The main objective of this book is to analyze illegal river-sand mining and its negative consequences in six riverfront Zhuang villages. By depicting the processes of the commercialization of river sand by diverse actors and their impacts on the river ecosystem, on disputes in the riverine communities of the Zhuang, and on sand governance, this study intends to provide new insights into this resource crisis. It addresses power relations, institutions, culture, beliefs, narratives, conflicts, and peace-building in several Zhuang communities in southwest rural China.

In order to achieve this objective, this book is guided by the following questions:

1) What drives the commodification of river sand in rural China?
2) How do competing rights of ownership of river sand affect its governance?
3) What is the role of the Zhuang communities and their ecological knowledge of river sand as a resource in the context of the river-sand crisis?
4) What are the sociocultural, economic, and ecological impacts of the ongoing market-oriented sand exploitation?
5) Which actors play a critical role in the commodification of river sand, and how does this multiplicity of actors create issues in the management of a "common-pool" resource among the six villages studied? This comprises questions such as the following: Who is monitoring the public river sand? Who is sanctioning the offenders? To whom are such cases reported, and what action should be taken regarding those who disobey? What forms does resistance against illegal river-sand mining take?

This book consists of eight chapters. Following this introduction, chapter 2 illustrates the field settings in the GZAR, then briefly addresses the fieldwork process and the methodology used during the undertaking of this research. It also explores the consequences of ineffective river-sand control and management as illegal miners emerged around the Maoling River to mine sand and sell it.

Chapter 3 goes on to examine how the boom in the sand market has shaped the commodification of river sand and its commodity chain. This chapter introduces the river-sand sellers, truck and ship owners, and river-sand markets that constitute the river-sand exploitation/value chain. The chapter furthermore presents the ecological consequences of illegal sand mining.

Chapter 4 seeks to provide answers to the question of what drives the commodification of river sand by first discussing how government policies and economic dynamics have impacted river sand use in the six villages. The chapter provides rich data on the financial aspects of house construction and infrastructural projects. Moreover, it introduces case studies to illustrate the problematic effects of infrastructural investment programs on the natural and social environment.

Chapter 5 presents the shortcomings of China's formal politics, both horizontal (agencies at the same level) and vertical (from the central government down to administrative villages). It demonstrates the role of politics and the approach to control on the part of the government at various levels with regard to resource management in rural China.

Chapter 6 focuses on river-sand mining from the perspective of the villages. The ownership debate and river-sand mining crisis are explored through case studies in the six Zhuang villages, where the statement that "river sand belongs to us" is heard in various forms, and involves not only the local communities, such as leaders of natural villages and cadres of the administrative village, but also the local government and its officials as well as the related laws. Rural Zhuang people who settle along the rivers for ease of access to water and river-based resources often possess these resources as communal collectives, and are thus strongly linked to them in their diverse sociocultural ways of life (e.g., for their livelihood, for group formation and belonging, and as avenues for spiritual nourishment). Nevertheless, the participation of villagers in decision-making and policy implementation with regard to commercialized river-sand mining is limited.

Chapter 7 zooms in on the sociocultural impacts and outcomes of sand marketization. Conflicts emerge within and beyond villages when villagers experience the loss of public river sand and feel unjustly treated with regard to profit-sharing. They have been suffering from the overmining of river sand in their territory. The chapter describes various forms of local resistance. The findings show that lineage customs and the public worship system are ways to reconstruct peace within the village. But the trust between villagers and government at various levels has been damaged due to the mismanagement of sand mining and its negative consequences.

The conclusion summarizes the findings and the contributions of this book. Taking a political ecology approach, the conclusion discusses how the transformation of law and policy with regard to river-sand mining reflects successes and failures in solving the sand crisis in southwest China. It relates to existing scholarly literature about environmental and natural resource management in current China and concludes that informal institutions, customs, and moral discourse do not have sufficient power to counter local illegal sand mining driven by the increasing national demand for river sand.

CHAPTER 2

The Research Setting and Conducting Research "at Home"

This chapter describes the setting, fieldwork process, and methodology of the research recounted in this book. I conducted thirteen months of multi-sited ethnographic fieldwork in six Zhuang communities located in the administrative area of Qinzhou City,[1] surrounding the Maoling River, the largest river in this city.

This book is an ethnography of a river and river resources. Wagner, Jacka, et al. (2018, 4–5) proposed to take rivers as ethnographic subjects instead of objects, because "rivers have social lives and therefore merit thorough ethnographic investigation and description." (3) Following their approach, this book explores the interaction between the river and the riverfront peoples, which is crucial to understand the historical transformation of the local environment. I used qualitative as well as quantitative methods to explore the consequences of ineffective river-sand control and management, as illegal miners have emerged around the Maoling River who mine sand for sale. In the following, I will first introduce the research area: the GZAR, and more specifically the six sampled villages along the Maoling River, located in Xintang Town and Huangwutun Town in the city of Qinzhou.

1 The Guangxi Zhuang Autonomous Region: Human–Environment Relations in Historical Perspective

The Guangxi Zhuang Autonomous Region covers an area of 237,600 square kilometers and administers the Beibu Gulf area of approximately 40,000 square kilometers (GBSYC 2018). It is in the subtropical monsoon climate zone, with an average annual temperature of 17.5 to 23.5° C, average annual rainfall of 841.2 to 3,387.5 millimeters, and average annual sunshine of 1,213 to 2,135.2 hours. The GZAR borders Guangdong Province to the east and Beibu Gulf in the

1 In the Chinese administrative system, certain sets of villages form part of specific cities and are subordinated to them in terms of regional administration and governance. I will explain the rural and regional governance structures and levels in more detail below.

 DOI:10.1163/9789004505919_003

south, and faces Hainan Province across the sea. It borders Yunnan Province to the west, Guizhou Province to the northwest, Hunan Province to the northeast, and the Socialist Republic of Vietnam to the southwest (Guangxi Government 2018). Map 1 shows the geographical location of the GZAR in Southwest China. Its capital city is Nanning.

The GZAR and Guangdong Province have historically been called Lingnan. I also discuss the Lingnan region here in order to compare my findings in Guangxi to a larger area and to show the similarities between the illegal river-sand mining and management crisis in Guanxi and that in Guangdong Province.

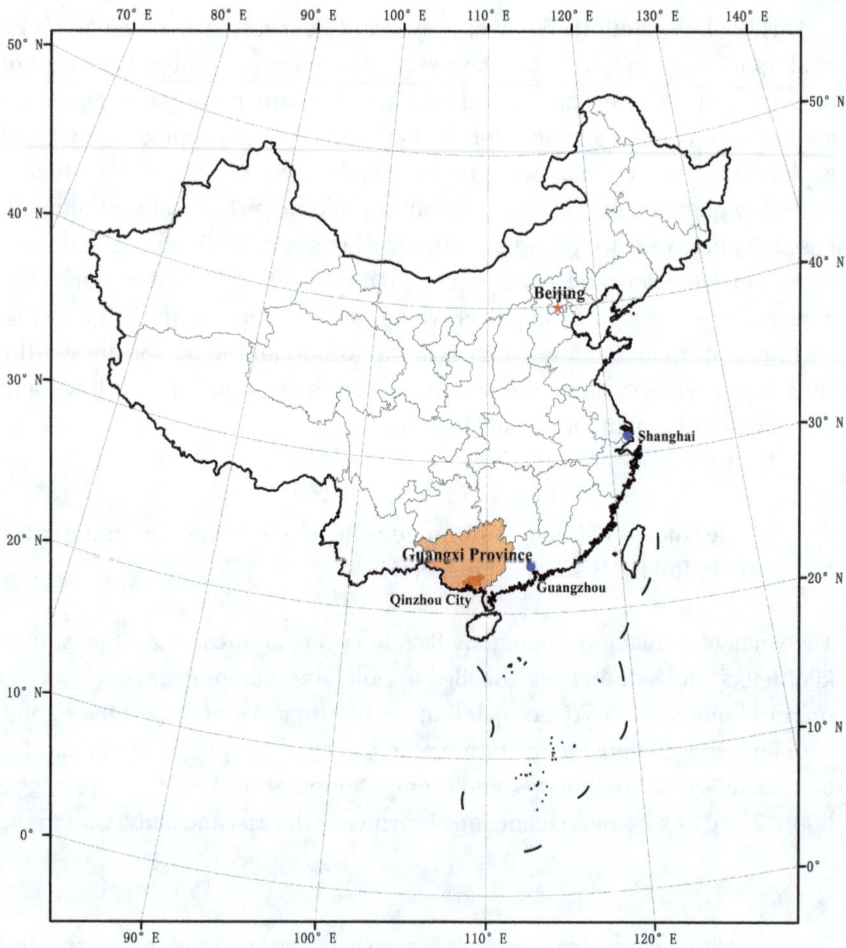

MAP 1 The location of Guangxi Province
CARTOGRAPHER: KE ZHANG, 2016

According to the data of the sixth national census of the People's Republic of China (PRC), carried out at the end of 2010, the Zhuang numbered 16.9 million, accounting for 36.7% of the total population of the GZAR (COSC and DPESNBS 2012). Other Zhuang reside in Yunnan, Guangdong, Guizhou, and Hunan Provinces (Kaup 2000, 33). Although the GZAR is a multi-ethnic region, it is the main settlement area of the Zhuang. The autonomous region offers advantages for the Zhuang:

> This administrative structure with elements of self-government makes it possible for different levels of the government to support ethnic minorities by lowering taxes and increasing public expenditures. Further, in minority areas, having minority status makes it easier to become a cadre (that is, a village head, village party secretary, or village accountant).
>
> GUSTAFSSON and SAI 2009, 193–194

From the state's perspective, autonomous regions were intended to assist national security and to facilitate the progress of all the national ethnic groups toward socialism (Harrell 1995, 24). The year 2018 marked the sixtieth anniversary of the GZAR. However, the Zhuang culture is much older than sixty years.

According to archaeological discoveries, there were primitive human races living in the area of the GZAR as early as 800,000 years ago. About 20,000 to 10,000 years ago, people living here learned to make and use sharpened stone tools. Between 10,000 and 6,000 years ago, the ancient people in the territory gradually moved out of the caves and river valleys and began living on the plains and coastal areas and formed tribes (MFAPRC 2018). According to Zhang (1997) and Qin (1998), ethnic groups who live in the GZAR today (including the Zhuang) are descended from these tribes. One of the old names of the Zhuang ethnic group is Luo Yue (骆越部族/ 雒越), which means "a paddy field in the mountains" (Zhang 1997; Qin 1997). Marks (1998, 54–55), discussing the whole Lingnan region, explains:

> In the lowlands of the river valleys and along the coast were various Tai peoples, the largest group [among them is] now called the Zhuang. … The lowland Tai peoples grew wet rice, cast bronze, wove silk, and by about 300 BCE had organized themselves into a state called Yue. With a state system and knowledge of sericulture and metallurgy, the Tai were the only people whom the Chinese (the Han) did not consider barbarians.

As discussed in the introductory chapter, the Zhuang have a rich and colorful culture that brought forth thousands of songs and dances, Zhuang brocade (one of the most highly valued silks and textiles, which became a tribute delivered to the emperor in the Song Dynasty), a specific clothing style, a rich food culture, and musical instruments (MFAPRC 2018).

The Zhuang also have a language of their own, which belongs to the Zhuang-Dai branch of Zhuang-Dong Austronesian, part of the Sino-Tibetan language family. The language is divided into two dialects (so-called Southern Zhuang and Northern Zhuang), although there are few differences between them (GBSYC 2015b). Chinese characters were used before 1955 for writing. However, a writing system based on the Latin alphabet was created in 1957 with the help of the Chinese government (Liang 2000). In their language, "the Zhuang did not refer to themselves as Zhuang, but as 'Bu duo.' They referred to the Han[2] as 'guest people'" (Kaup 2000, 46). Nowadays, however, most of the schools exclusively use Mandarin for teaching, and Mandarin is overwhelmingly used in the media. Therefore, the children in rural areas can also speak Mandarin.

The Zhuang were recognized as an ethnic group by the governors of China mainly from the Song Dynasty (960–1279) onward (Took 2005, 7), when there was a rebellion by Zhuang leaders including Nong Zhigao (1025–1055) (X.-F. Huang 1962). The political system in this area was originally defined by the so-called Tusi (土司) system (chieftaincy; literally, "native superintendency"), in which the officers of the Tusi administration were Zhuang (Li and Pang 2000, 194). The last native chieftaincy administration in Guangxi ended in April 1928 during the political reform of the bureaucratization of native offices (*gaitu-guiliu* 改土归流), when the government abolished the Tusi system and installed its own officials in place of Tusi officers (Took 2005; Zhang 1997, 657–676).

As the Chinese Empire expanded, the transformative impact of civilizational forces reached the Zhuang, including the establishment of schools, the propagation of the ethics of feudalism, the transmission of iron-working and cattle-farming techniques, the construction of water management, and the introduction of farmland irrigation (GBSYC 2015a). The economy in the Lingnan came to depend on rice farming, making use of the rich river system in this area, which includes the Red River, Left River, Right River, West River, Liu River, Rong River, and Maoling River. Lingnan's agricultural economy has played a significant role in shaping the local environment. According to Marks (1996, 53), "Late imperial Lingnan was dominated by Han Chinese, and it was the Chinese who mostly remade the natural environment into one that

2 The Han people moved from Zhongyuan to Guangxi in large numbers following the Qin and Han Dynasties. According to statistics from 2007, the population of Han people in Guangxi was 31 million, accounting for 62% of the total population, mainly distributed in the southeast of Guangxi (http://www.gxzf.gov.cn/mlgxi/gxrw/ftrq/t1003598.shtml).

supported the settled agricultural practices and cultures commonly understood as Chinese: irrigated paddy fields from which two or more crops of rice could be harvested from the same plot of land with little or no fallow period." Marks's work (1996; 1998, published in Chinese in 2004) offers a good interpretation of the interaction between nature and the people in Lingnan . Marks (1998, 53) explored why "the Han people had come to occupy the richest farmland in the river valleys and the Pearl River Delta, while the non-Chinese held lands in the hills of northern and eastern Guangdong and in the western half of Guangxi" by the late imperial period.

Significant for my study, Marks (1996, 1998) also explored the river system's role in shaping people's livelihoods and the transformation of nature in Lingnan. He argued that the function of many cross-border rivers was to transport silt, which shaped river deltas in Guangdong Province, including in areas where the Zhuang lived. Marks (1998, 66) reported that "silt carried downstream in the West, North, and East Rivers had been setting out in the bay [the Pearl River Delta], slowly creating the upper reaches of the delta." The fertile plains and valleys on the upper courses of rivers where the Zhuang had settled attracted other ethnic groups to the river valleys—in particular the Han people. Over time, cities such as Guangzhou City grew enormously because of this area's agriculturally rich capacity, and through practices of water control and irrigation (Marks 1998, 63–65). Furthermore, rice and other crops became cash crops because of the growth of cities and global trade connections (such as the trade in silk and tea). According to Marks (1996, 60), "The most immediate stimulus for the commercialization of the economy was the new demand for Chinese goods—in particular silk and porcelain—from European traders who arrived in the South China Sea in the sixteenth century."

The growing silk industry resulted in the commercialization of the major crops and the transformation of land-use patterns. Marks (1996, 65–67) estimates that "commercialized agriculture, even in the early decades of the eighteenth century, may have claimed half or more of the arable land in Guangdong Province." This resulted in the expansion of the "mulberry tree and fish pond system" (*sangjiyutang* 桑基鱼塘) (61):

> As peasant farmers dug up the rice fields for the "mulberry tree and fish pond" system, they turned to the market to purchase their food, and markets grew both in size and in number. ... By the middle of the eighteenth century, the market system linked the various parts of Lingnan into a single agroecosystem. That being the case, to look at farming practices in just one part of the system misses the larger picture: the system as a whole was not sustainable without greater and greater inputs from

> outside. Guangxi Province was drained to sustain the Pearl River delta. (61, 76)

Significantly, Zhang (1997) argued that the development of international trade in the Qing Dynasty (1644–1911) greatly impacted the self-sufficient agricultural livelihoods in Guangxi. The Guangxi region imported more and more goods, mainly from Europe. After the First Opium War (1839–1842), opium and other Western goods were rapidly brought into the region by Western countries. These goods were transported from trading ports to the interior via boats on the rivers. The trade resulted in the outflow of silver, the bankruptcy of farmers, the concentration of land, and the demise of the domestic silk industry (Zhang 1997, 820–825; Li and Pang 2000).

In particular, the development of technology in modern silk manufacturing in Western society and Japan greatly contributed to the decline of the rural silk industry in Guangxi. A similar situation was observed by Fei (1939, 16–17) in the Yangtze River area:

> As Japanese silk was dumped on the American market in the latter part of the year [1934] China's silk export dropped to the lowest point accordingly. The volume of silk exported amounted to only one-fifth of that in 1930, a fact which is indicative of the depression of the Chinese silk trade. … The internal market for silk has shrunk at the same time due to the same forces of industrial revolution affecting the weaving industry. The consequence of the shrinkage of the market is the break-down of the traditional domestic silk industry in the rural industry.

Nevertheless, the Zhuang area continued to maintain its agricultural system, even in the face of the depression of the Chinese silk trade. The majority of people still conducted agriculture. Kaup (2000, 29–30) states:

> The harsh mountainous terrain in China's southwest has historically kept the area even more removed from the central government than has its actual distance from the capital. Of Guangxi's more than 230,000 km^2, over 70 percent is covered with mountains and 20 percent with rivers, leaving only 10 percent suitable for farming. … Seventy-two percent of Guangxi's agricultural fields are still rice paddies, however, and rice accounts for over 80 percent of Guangxi's total food production.

A survey of more than 900 minority villages located in 22 provinces in 2002 shows that

> minority villages located in southwest China clearly have a poorer economic situation than the average majority village. … Location is found to be the single most important circumstance working against a favourable economic situation for Uyghur, and particular for Zhuang, Miao, Yi, and other southwestern minority villages. We also find that a low village income induces many ethnic minority persons to be involved in long-distance migration, but for some minorities, this potential source of economic development is thwarted by their ethnicity.
>
> GUSTAFSSON and SAI 2009, 194

Even though there is little farmland available, the Zhuang people living here have heavily relied on farming as their primary source of income. The Guangxi local chronicles recorded how the diffusion of agricultural skills improved the local Zhuang peasant farmers' grain harvests (Wang 1988; Li and Pang 2000). A similar narrative is told by the Chinese Communist Party (CCP) and the central government:

> The Zhuang area is basically agricultural, but before 1949 the local people never had enough to eat despite their hard work and the favorable natural conditions. By 1983, they had raised grain output by 158 percent thanks to improved field management and the 500,000 water conservancy projects built since liberation.
>
> MFAPRC 2018

This shows the achievements of the CCP in increasing the Zhuang people's harvest. The construction of waterworks has raised the capacity to control water in order to avoid flooding and drought. The land reclamation policies have contributed to the increase in the amount of land under cultivation under the CCP's agricultural policies since the 1950s. However, these policies have profoundly changed agricultural practices and the ecological situation in the Zhuang region. For instance, land reclamation also resulted in deforestation, which the government has tried to address (Wang 1988, 519–520).

2 Qinzhou City, Zhuang Towns, the River, and the Villages

Qinzhou City is located along the coast of southwest Guangxi. According to the Qinzhou government's geographic census in 2017, the population was around 4.1 million (QZGOV 2017). Qinzhou City is home to members of twenty-three ethnic groups including the Zhuang, Yao, Miao, Jing, Yi, Li, Maonan, Tujia, and

Han. The Zhuang group, which developed from the Luo Yue people, is the oldest ethnic group in Qinzhou (QLCCC 2018, 2144–2145). The city's ethnic minority population (all ethnic groups excluding the Han) is about 413,000, making up 10.24% of the total population, of which the Zhuang number 406,000, accounting for 10.06% of the total population. There are 943 Zhuang-inhabited natural villages (communities) within the administrative area called Qinzhou City. These communities are distributed over 116 administrative villages in 16 different towns of Qinbei District, Qinnan District, and Lingshan County, all part of Qinzhou City. The Zhuang population accounts for more than 50 percent of the population in the six towns of Dazhi, Guitai, Dasi, Nameng, Xintang (in Qinbei District), and Huangwutun (in Qinnan District) (QZGOV 2017; QLCCC 2018, 2144–2145). For example, Xintang Town has 50,762 residents, including 48,899 Zhuang, 46 Miao, and 23 residents from other minority groups. There are also 1,789 Han residents (XTGS 2015).

According to the "Minority Autonomous Township Administrative Work Rules" of 1993, "a township can apply for Minority Autonomous Township status if the percentage of ethnic minority households accounts for over 30% of the total population in the jurisdiction" (quoted from Gustafsson and Sai 2009, 195). Huangwutun Town and Xintang Town, the two field sites for this study, are both Zhuang towns because they have over 50 percent Zhuang residents (see Map 2).

2.1 *Huangwutun Town*

Huangwutun Town is located in the northwest of Qinnan District and four kilometers from Qinzhou City. It consists of 156 natural villages, 16 administrative villages, and one urban community, with an administrative area of 226.7 square kilometers. The main languages spoken in the township are Zhuang, Hakka, and Cantonese. Peasants living there grow commercial crops such as rice, sweet potatoes, corn, peanuts, soybeans, vegetables, and fruit (QLCCC 2018, 70; Li and Pang 2000, 102). Huangwutun Town had 55,161 residents in 2008; 51,731 had agricultural residency status and 3,430 had non-agricultural residency status.[3] The minority population was 38,723, most of them Zhuang, accounting for 70.2% of the population in 2008 (QLCCC 2012, 282).

3 Residency status in China is determined through the *hukou* or household registration system. Lei and Li (2012, 1) explain: "Beginning in the 1950s, the Chinese government officially promulgated its *hukou* system to differentiate residential groups. … There is a dual classification in a person's *hukou*. The first is one's residence, which is commonly referred to as rural/urban area; and the second is one's socio-economic eligibility, which is commonly referred to as agricultural/non-agricultural category [*sic*]. Individuals registered under the agricultural category depended mainly on their own labor and the fluctuating harvests for survival;

MAP 2 Qinzhou City
CARTOGRAPHER: KE ZHANG, 2016

In 2008, the total industrial output value of Huangwutun Town was 733 million yuan; the total output value of agriculture, forestry, animal husbandry, and

individuals registered under the non-agricultural category, on the other hand, were entitled to a 'cradle-to-grave' welfare package provided by the government. As a result, urban residents were seen as superior to rural residents in terms of socio-economic status." Chan (2014, 1) adds that the reform of the *hukou* system since 1980 "means allowing migrant workers to gain permanent residency in Chinese cities, and thus to be covered by the urban public service system in their place of work. In effect, China's migrants need to become full urban residents."

fishery was 212 million yuan; the financial income was 12.06 million yuan; and the per capita income of farmers was 3,980 yuan (QLCCC 2012, 282). In 2019, the industrial output value of the town was 559 million yuan, the financial revenue was 26.43 million yuan, and the per capita income of residents had risen to 14,627 yuan. Due to poverty alleviation projects, only 0.2% of the total town population was categorized as poor in January 2020 (AGWRHT 2020).

The fishing industry is mainly based along the Maoling River and its tributaries. The stretch of the Maoling that passes through Huangwutun Town alternates between salt and fresh water due to the fluctuation of the sea tide. Fish and shrimp are plentiful and their meat is fresh and tender. The town had the highest catch of freshwater fish and shrimp in Qinzhou City in 1990 (Li and Pang 2000, 102). In addition, the town had 2,601 hectares of arable land in 2008, consisting of 2,238 hectares of rice paddies and 363 hectares of dry land, as well as 67 sand-mining sites (QLCCC 2012, 282–284).

The government implemented several developmental and infrastructural projects in Huangwutun Town at the time of my fieldwork. For instance, the expansion of the Huangwutun Bridge, which had been damaged by illegal sand dredgers and overloaded trucks, was started in September 2016, and construction was finished in April 2018. A total amount of 5.7 million yuan was invested by the town's renovation project, which covered 70 percent of the project's costs. It related to the bridge project, and a cumulative investment of 4.8 million yuan enabled the completion of the construction of a so-called "eco-village construction" (*shengtai xiangcun jianshe* 生态乡村建设); this project included the installation of 84 solar street lights and the construction of a sewage treatment plant, two major drains leading to transfer stations, and paved roads in three villages.

In order to eliminate poverty at the roots, a poverty alleviation assistance program was established and supporting measures were put in place. In April and September 2016, two clusters of poverty alleviation projects were started with a total investment of 350,000 yuan and delivered 17,400 chickens, more than 200 pigs, 260 ducks, and 14,000 kg of feed to poor households in Huangwutun (AGWRHT 2017). With these measures, the township government implemented projects that the higher government had planned in advance.

My fieldwork data from Huangwutun Town shows that the town had a rather high emigration rate. In Naxia Village, a whole branch of the village's lineage (one-fourth of a total of 500 villagers) emigrated to Qinzhou City, while many households of Nazhong Village worked and lived in Nanning City. During the Qingming festival, many households and individuals came back, worshipped at their ancestors' graves, and stayed for two days in the village. These households were unaware of what had actually happened in the village while they were away (see section 4). Some of the returning emigrants answered "I don't know" to my questions about sand mining in Naxia Village.

2.2 *Xintang Town*

Xintang Town is located on the northern border of Qinzhou City. The township is 58 kilometers from Qinzhou City and 70 kilometers from Nanning City. It is governed by Qinbei District. It has eight administrative village committees and one neighborhood committee, and a total of 65 natural villages. Xintang's administrative area covers 116 square kilometers. Zhuang, Xinli, and Cantonese are languages spoken in the township. The main crops produced there are rice, sweet potatoes, corn, sugar cane, and silkworm mulberry. Lychees, bananas, and Chinese yams are also popular in Xintang (QLCCC 2018, 75; Li and Pang 2000, 106). In 2008, Xintang Town had 44,887 residents, of whom 43,809 held agricultural residency status. The Zhuang population accounted for 96.4% of the town's total population (QLCCC 2012, 434).

In 2008,[4] the total output value of Xintang's agriculture, forestry, animal husbandry, and fishery were 212 million yuan, the financial income was 11.6 million yuan, and the per capita income of farmers was 3,675 yuan (QLCCC 2012, 434). The town had 1,607 hectares of arable land, including 1,153 hectares of rice paddies (434). The farmers' annual income per capita in Xintang Town had increased to 7,703 yuan by 2013 (LGCXT 2013). In 2019, Xintang had a total output value of 510 million yuan in agriculture, animal husbandry, and fishery, while the tax revenue was 4.29 million yuan. The net per capita income of residents was 14,068 yuan (AGWRXT 2020).

Xintang Town has transformed from a peasant economy to a (migratory) labor economy. Essentially, many peasants in Xintang Town have become migrant workers who work outside the town while trying to maintain their farmland in the village. Table 1 shows the number of non-peasant and migrant workers as well as the total number of households and persons registered in Xintang Township. The categories were established by the township government, which classified one-third of all laborers as migrant workers. This figure is somewhat misleading because the migrant workers were also counted among the resident population of resident households. Even though the residents were moving in and out, they were officially recorded as permanent residents by the township government. In other words, many of the agricultural

4 There is no record of the total industrial output value of Xintang Town because it is an agricultural town.

laborers were actually migrant workers who frequently returned during the year to work on farms (LGCXT 2013).

The Qinzhou government implemented the Unified Urban and Rural Household Registration System (the *hukou* system) on September 1, 2014. The new policy allowed all residents to claim the same residence status regardless of the original location of their *hukou* registration. In Qinzhou City, many communities contain many immigrants because more and more migrant workers have settled down in urban areas. Statistics collected by the Xintang government in 2014 showed that 173 people immigrated to Xintang Town, while there were 1,622 residents who emigrated, in both cases mostly within Guangxi Province.

2.3 *Six Villages along the Maoling River*

Some scholars who focus on rivers explore the relationship between rivers and peoples and highlight the crucial role of rivers for local people's life and well-being (Wagner and Jacka 2018; Strang 2014; Krause 2010; Weir 2009; Barnes 2014). Several studies have more specifically discussed the relations between rivers, local economies, and the environment in rural China. Skinner (1977) conducted his doctoral research on the structure of markets in the Chengdu Plain in China in 1949. He highlighted the significance of the river system in forming the regional economies and livelihoods:

> Taking the river basin as the essential regional determinant is particularly appropriate in the case of agrarian China, where crop inventories and productive technique were specially adapted to a plains-and-valley ecology and where water transport was of the greatest importance. In most Chinese physiographic regions, the river system provided the skeleton of the transport network that underlay the region's functional integration. (282)

Marks (1998, 2012) is another scholar who has highlighted the crucial role of rivers in forming the regional eco-culture system, as discussed above.

This book focuses mainly on the Maoling River, which is the largest river in Qinzhou City; its source is in Bancheng Town, and it flows through Xintang Town, Huangwutun Town, and several other towns in Qinzhou City. It eventually flows into the Maowei Sea in the south. Mei Village,[5] Nayang Village, and Na Village are located upstream along the Maoling River, while Naxia Village,

5 The book uses pseudonyms for the villages and those who participated in the research.

TABLE 1 Labor statistics of Xintang Town

	Agricultural laborers			Migrant workers	Migrant workers outside Guangxi Province	Resident households	Resident population
	Total	**Male**	**Female**				
Total	30,348	16,707	13,641	10,351	7,557	10,807	49,732

SOURCE: XINTANG TOWNSHIP GOVERNMENT STATISTICS, FEBRUARY 2013

Nachun Village, and Nazhong Village are located downstream. The population distribution of Qinzhou City is mainly concentrated along the city's three major rivers, and the Maoling River Basin has the largest population, accounting for about half of the total population (Li and Pang 2000, 199). Of the six Zhuang towns in Qinzhou City, four of them are located along the Maoling River, including Xintang Town and Huangwutun Town (146, 194).

The total length of the main river is 112 kilometers, and the basin area covers 2,959 square kilometers. The basin's average elevation is 109 meters above sea level. The river passes through Qinzhou City and also through the nearby Fangchenggang City. The normal depth of the Maoling River is one to four meters, while the normal width of the river is between 120 meters upstream and 300 meters downstream. It has a sandy riverbed. The tide from the Maowei Sea can be traced upstream to Huangwutun Town's Leather Dam. Thus, the river is a tidal river downstream of the Leather Dam. River water is abundant, according to many years of observation by the Huangwutun Hydrological Station. The average annual flow is 82.12 cubic meters per second, the average annual runoff is 25.9 billion cubic meters, and the annual runoff depth is 1,000 mm. Due to the natural variation in precipitation, the river flow changes greatly during the year. During the flood season between April and September, the average runoff is 1,999 million cubic meters, accounting for 77.2 percent of the annual runoff. The Maoling River hosts diverse fish species and plays a significant role as a means of water transport. It also provides water for agriculture and drinking, as well as industrial purposes (Li and Pang 2000, 146–148). The land along the river basin is fertile and it is the main area for food production. However, because of the narrow riverbed and the gentle slope of the lower reaches of the Maoling River, when a flood coincides with tide inversion, it causes disaster downstream of the Leather Dam (199).

The volume of sand transported by the Maoling River each year is 553,000 tons, which is more than any of Qinzhou's other rivers. River sand is mostly yellow-white fine sand, suitable for building materials (Li and Pang 2000, 175). Since the 1990s, the sand of the Maoling River has been harvested, in particular through illegal sand mining and sale, and numerous deals have been struck along the banks of the river. According to government officials in Huangwutun Town and Xintang Town, those mining sites, as well as sand dredgers on the river itself, lack mining or business licenses. Streamlined sand mining and effective sand governance are absent in the region. At the same time, the price of sand used for construction has shot up from several yuan per cubic meter to more than one hundred yuan per cubic meter over the last two decades. As a result, the violators have stolen many of the river's resources and the Maoling River is undergoing rapid transformation. I adapted a satellite image to show the location of sites for storing and selling sand along the Maoling River in Huangwutun Town (see Figure 2).

In recent years, the sand of the Maoling River has been extracted in large amounts, and many disputes and conflicts concerning sand have emerged along the river. Many villages, local governments, and private sand companies are involved in these disputes. My intention was to capture and render the bigger picture of illegal sand mining in Qinzhou City. Therefore, I eventually chose six villages within Xintang Town and Huangwutun Town as research sites. Xintang Town is located upstream along the Maoling River, while Huangwutun Town is situated downstream. From each town, three villages were chosen. The three villages selected in each town are located close to one another, and thus decisions concerning sand-mining activities and management in any one of the villages have an impact on two nearby villages. In addition to conducting multi-sited research within the six villages, I expanded the scope of the investigation by considering what happens at different levels of governance. Before continuing, it is necessary to lay out the complex levels of rural governance (Table 2).

In this administrative system, counties and districts form part of the city while each town belongs to a district or a county. The town encompasses the administrative villages and urban neighborhoods. The administrative village is constituted by what can be translated as "natural" villages (*zirancun* 自然村), while a natural village can be further subdivided into several production teams. Natural villages can be understood as settlements and territories which are self-governed through particular lineages, village heads, captains of production teams, and elders. An administrative village has a fixed governance

FIGURE 2 Illegal sites for storing and selling sand in Huangwutun Town
SOURCE: GOOGLE MAPS, LAT. 21.981065°, LON. 108.523119°, JULY 20, 2017

TABLE 2 Governmental structure of the research sites in Qinzhou City

District (county)	Town	Administrative villages	Natural villages	Location
Qinbei District Government	Xintang Town	Nagan Administrative Village	Mei Village Na Village Nayang Village	Upstream along the Maoling River
Qinnan District Government	Huangwutun Town	Naqiu Administrative Village Nadong Administrative Village	Naxia Village Nachun Village Nazhong Village	Downstream along the Maoling River

SOURCE: FIELD DATA, 2017

structure: a village committee with three to seven elected cadres and a CCP branch with one secretary. The Organic Law of the Village Committees of the People's Republic of China (OLVCPRC87, OLVCPRC98, and OLVCPRC10) has been implemented since the late 1980s. It prescribes that villagers can elect their own leaders to practice grassroots democracy. Based on this law, the village committee is constituted by three to seven villagers including a director, deputy directors, and committee members. The tenure of a village committee is five years. Apart from the elected village committee, there are branches of the CCP at all levels of the village governance structure. The party agency is the most powerful unit inside the government, by law. Indeed, there has been much criticism of the realities of the implementation of the Organic Law of the Villagers Committees (Feng 2009) and of authoritarian control over village elections in China (Luo 2014).

Map 3 shows the location of Mei, Na, and Nayang Villages. It also displays Yang Village, which joined Mei and Na Villages in taking action against illegal sand mining in 2018 (see chapter 7). Furthermore, Map 3 displays Naluo Village, the home of the ancestors of Mei Village, showing the lineage connections between these villages.

My home village, Mei Village, became my central research site, because my focus on the sand-mining situation originated there.

TABLE 3 The rural power structure and the population of the six Zhuang villages

Level	Demography	Structure
Administrative villages	a. Nagan Administrative Village has 1,450 households, 7,229 villagers, and 85 CCP members. b. Naqiu Administrative Village has 910 households, 4,159 villagers, and 66 CCP members. c. Nadong Administrative Village has 1,368 households, 5,648 villagers, and 73 CCP members.	Nagan Administrative Village has 10 natural villages and 20 village groups. Naqiu Administrative Village has 8 natural villages and 14 village groups. Nadong Administrative Village has 11 natural villages and 20 village groups.
Natural villages	a. Mei Village has 505 villagers. b. Na Village has 1,200 villagers. c. Nayang Village has 1,300 villagers. d. Naxia Village has 290 villagers. e. Nazhong Village has 11,000 villagers. f. The sixth production group of Nachun Village has 450 villagers.	Mei, Na, and Nayang Villages are subordinate to Nagan Administrative Village. Nazhong Village is subordinate to Nadong Administrative Village. Nachun Village and Naxia Village are subordinate to Naqiu Administrative Village.

SOURCE: FIELD DATA COLLECTED FROM THE THREE VILLAGE COMMITTEES AND THE NATURAL VILLAGES IN 2015–2017

2.3.1 Mei Village: An Example of Agricultural Crisis and Its Impact on Outmigration

When I started my research in Mei Village, I learned that it had 70 residents when the Communist Party took over the government of China in 1949, and 505 residents by 2014. The Household Responsibility System (*Jiating Lianchan Chengbao Zerenzhi* 家庭联产承包责任制), which allows individual households to contract to use a specific plot of land while collective landownership is retained by the village, was implemented at the end of the 1970s, and 235

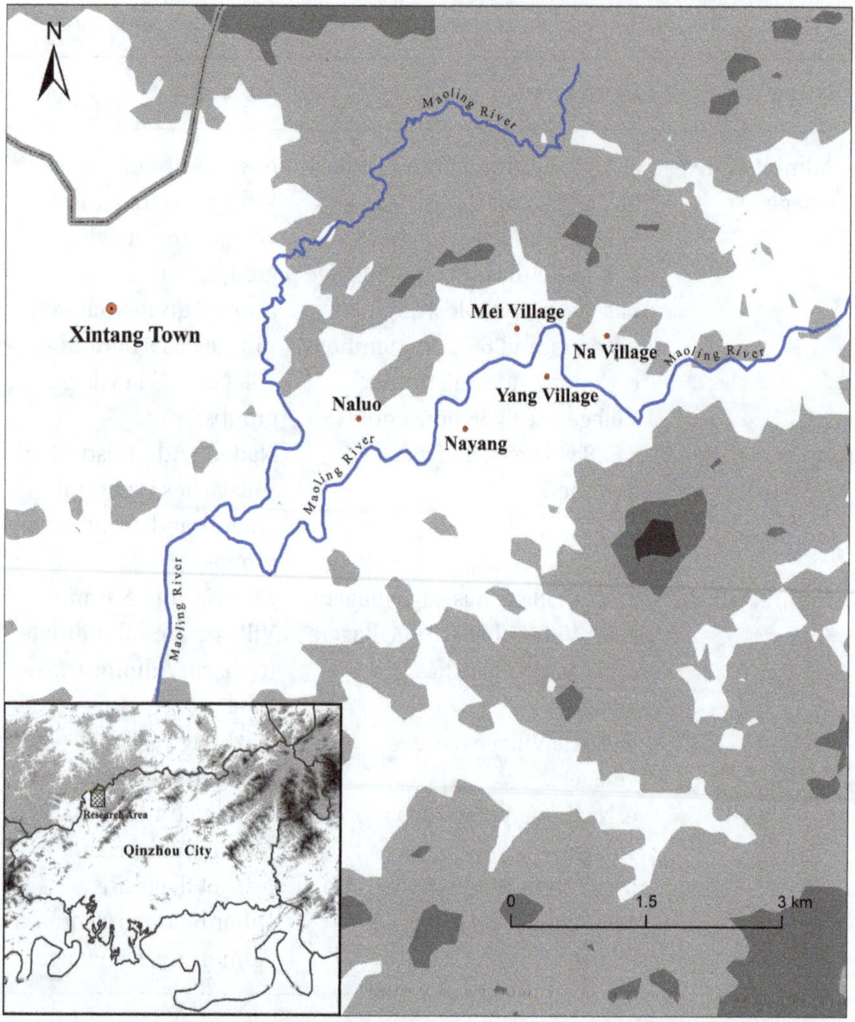

MAP 3 Villages upstream of Maoling River
SOURCE: CARTOGRAPHER KE ZHANG, 2016

members of Mei Village signed the contract (*chengbao hetong* 承包合同). They joined the Production Brigade of Nali for a term of 15 years (January 1, 1985–December 31, 1999) on November 17, 1984. Every adult was granted 0.683 *mu* of farmland (455 m^2) for this period of time. The contract was renewed in 1994 and expires on December 31, 2029.

The total land area of Mei Village was about 148.3 *mu* (9.8 hectometers) in 1952. Because the amount of available land was limited, the income of village farmers was relatively low, because it was generated through the cultivation of paddy rice and lychees and poultry breeding.

In recent years, however, income from agriculture has not been the main source of income for households in Mei Village because of the increase in migrant labor, which plays a significant role in the village economy. Some farmers informed me that there is barely any profit from farming, but growing rice and vegetables, as well as fruit, is still significant for a family for self-subsistence. Interviews I carried out in this village in 2014 indicated that some households would earn only 10,000 yuan per year by farming, while others might obtain even less. For instance, one peasant planted 0.8 *mu* (533 m^2) of rice paddy in 2014, from which he obtained 800 *jin* (400 kilograms) of rice and received a subsidy of 300 yuan from the government (Figure 3).

FIGURE 3 A farmer watering his paddy fields in Mei Village
SOURCE: FIELD DATA, 2015

The peasants of Mei Village were used to planting two crops of rice each year. My survey showed that 44 households had stopped planting the second crop by 2014. Lack of manpower is one of the major reasons why people have stopped planting late rice; this is particularly the case when both parents of a household are working outside the village.

Sugar cane cultivation was encouraged by the government by offering financial aid. The state paid 460 yuan for 500 kg of sugar cane in 2013, and 360 yuan in 2014 (see Figure 4). This was added to the price farmers received when selling the sugar cane. Two households in Mei Village had a good cane harvest in 2014. One household earned 8,000 yuan and the second household more than 10,000 yuan. Moreover, there were more than five other households that gained around 3,000 yuan from sugar cane.

While the out-migration of labor was the primary cause of the agricultural crisis, it was aggravated by consistently low prices for rice and fruit. "A good harvest does not mean a better income," the peasants said. In 2018, there was a great harvest of lychees in Xintang Town. However, the price of lychees was rather low in the region. Pictures and videos of the peasants dropping their lychees onto the roads and destroying the fruit by stamping on them were posted on social media and reported on government websites. The peasants destroyed their own crop because the only businessmen who wanted to buy their lychees offered the incredibly low price of 0.5 yuan per 500g. This situation hurt the peasants deeply. Mei Village is in this regard exemplary of many villages across China.

In the same year, the central government took note of the plight of village farmers. On September 23, 2018 (the autumnal equinox in the lunar calendar), the Chinese president initiated the first "Chinese Farmers Harvest Festival" (*zhongguo nongmin fengshoujie* 中国农民丰收节)(Y. Liang 2018). This step shows that the central government realized that there is a crisis in agriculture, and an associated threat to food security. Attaching great importance to agriculture forms the foundation for social stability and plays a key role in governance (Y. Liang 2018).

2.3.2 Migrant Workers Who Settle in Urban Areas

Following Deng Xiaoping's market-oriented reform in the late 1970s, rural China has experienced a fundamental transformation in terms of people's livelihoods and lifestyles. Migrant workers moving out have become a common phenomenon in rural Xintang Town and Huangwutun Town (see also Zhang 2002). Residents of Mei Village have also increasingly moved to urban areas to look for employment or start businesses. The villagers have learned some new things (e.g., languages) during their working experiences in urban

FIGURE 4 Harvesting sugar cane and fishing
SOURCE: FIELD DATA, 2015

areas (Guo and Zhou 1996). They have become "rural-to-urban migrants" (Lei and Li 2012) and "new-generation migrant workers" (Z. Cheng 2014). The rural village, which was considered a "rural acquaintance society" (Fei 1998), has become a "semi-acquaintance society" (He 2013). Nonetheless, most ethnic minorities are less likely to migrate, and to migrate for a shorter time period if they do, compared to the Han majority (Gustafsson and Sai 2009, 201; Howell, Gustafsson, and Ding 2018, 1). Howell (2017, 209) concludes: "First, rural–urban migration boosts average rural income irrespective of the household's ethnic status, although considerable ethnic heterogeneity exists. Second, minority households tend to benefit less from rural–urban migration compared to Han households even after controlling for unobserved heterogeneity."

While conducting my surveys in the villages, I mostly met women, elderly persons, and children. Some grandparents were taking care of children while their parents were working outside the village. I met an older woman who was looking after the four children of one of her sons while her other two sons had taken their family with them to Nanning City. Mothers must balance home and work responsibilities. Some women must take care of not only children but also sick elders. Many men live a double life and are responsible for both taking care of the rural home and earning money outside the village.

During fieldwork I understood "migrant workers" to be those who regularly worked and lived outside the village for more than six months at a time. This was the case for about 61 percent of the 505 Mei villagers in 2014. Every household had members working in Guangdong Province or in Nanning City, Qinzhou City, or other towns within Guangxi Province. Middle-aged men would work as construction workers in Nanning City or nearby towns. The daily wage for a construction worker ranged from around 180 yuan to 300 yuan, while factory workers could earn more, namely between 3,500 and 6,000 yuan per month. In March 2015, an informant told me that his income as a construction worker peaked in 2013, at 60,000 yuan, while his entire income in 2014 was only 45,000 yuan. Usually it was easy to find a job on construction sites in Nanning City (90 minutes away from Mei Village by bus), where urbanization was taking place rapidly. Construction workers often commute between cities and villages. Some of them do so by riding motorcycles for 90 minutes or by traveling by bus for two hours between Nanning City and the village. Some of them live on construction sites in the city, sleeping for free on simple beds until the building has been finished and their salary paid. These migrant laborers told me that it was impossible to work on construction during the rainy season, and so many of them return home. Otherwise, they come back to their villages for weddings, funerals, and festivals, as well as to "send our children to school." A man told me, "I am very afraid of phone calls" when working on construction sites away from his village (Mei Village, October 20, 2015). As a father of two children, he needed to return home when the kids were sick as well as for funerals and various festivals in the village. For him, a phone call implied that "something happened" and he had to return home soon. He also explained that he could not work in a factory, because a factory worker must adhere to a stable working schedule, which he cannot do because of familial and social duties. "We have to be flexible because we need to deal with emergencies," he said, even though flexibility means fewer chances for a better job and lower wages. Translocality and flexibility are the main common characteristics of the male villagers' lives, as well as those of some females.

Some female villagers told me that they had been working for years in factories in Guangdong Province before they got married. They believed that they too could earn money for the family just like men. Some women pointed out that working outside the village broadened their horizons and gave them more financial independence. During my research period, from 2014 to 2016, some married women were working hard in the factories in Guangdong Province or nearby cities. As a result, the wealth differences apparent in the village are based on family labor. Commonly, when more household members work outside the village, the household is wealthier.

The migrant workers frequently return to Mei Village during their periods of work outside the village. Celebrating Lunar New Year is the most important reason for migrant workers to come home. In addition, they return to fulfil their sociocultural duties, such as worship, and to attend weddings or funerals. Some villagers confirmed that they return eleven times per year, while some just visit three times. These return trips have contributed to consolidating their social relationships (Zhu and Qin 2016). However, the frequency with which villagers return to their village is a rather individual matter and varies widely.

Some villagers gradually settle down in cities and towns. In particular, younger generations move to the cities because they hold a degree from a technical school or a university. Around twenty households from Mei Village moved to the cities and settled there in 2014, and twelve households bought apartments in cities and towns between 2015 and 2019. A rather new phenomenon was that some migrant households would take their children and elderly with them, and the children attended school in the cities. The mobility of migrant workers between urban and rural areas within Guangxi has contributed to the growth of this "rather new phenomenon." One retired teacher's three sons, for instance, have bought a total of four flats in Qinzhou City. The whole family moved to the city and left their village land to their relatives. They said that:

> Ninety percent of the households in our sons' communities are our people, who speak our language [Xinli]. We speak the same language; thus, we do not need to speak Cantonese or Mandarin at all. The owners of the grocery stores, and the people who sell vegetables and fruit, all speak our language.
>
> Interview with ZHU ZHIJI, his wife, and one of their sons, in their new apartment in Qinzhou City, August 2015

As I have stated, some households not only keep apartments in the cities, but have also built new houses in their home villages. They live in their rural homes during festivals or public worship activities, while they live in their urban apartments (whether bought or rented) during their working periods. Chinese governments at various levels have vigorously implemented rapid urbanization policies, and they have reformed the residence registering system to encourage more migrant workers to settle down in urban areas. They have built more schools and hospitals to provide those migrant workers with better public services. These policies have encouraged Mei residents to move out of the village and to settle down in cities and towns.

3 Starting Research in Mei Village: Topic Determination During Fieldwork

In order to gather further research ideas and to prepare a basic outline for my research project, I went back to my hometown—Mei Village, Xintang Town, Qinzhou City, Guangxi Zhuang Autonomous Region, China—for one and a half months of preliminary fieldwork in February and March 2014. I was born and grew up in Mei Village, and I am still an official resident of that village. During this first field trip, my attention was drawn most of all to changes in people's lifestyles and to socioeconomic changes, as shaped or induced by governmental infrastructure projects. A Chinese idiom, "earth-shaking changes" (*fantianfudi* 翻天覆地), was used to depict this recent transformation by the people I spoke to.

Some villagers mentioned the new roads, the morning market, the new toilets, and a garbage collection point. A few villagers brought up the issue of how to spend public money acquired by selling river sand. The people generally seemed enthusiastic about the new projects in the village. They also discussed the case in which a household had refused to contribute a small piece of its land for a planned basketball court unless the household members received compensation. Several other landowners had sacrificed their land for the public benefit so that the village could have infrastructure for sports. I noticed how the village's infrastructure had developed since I left in 2012. The road from the township to Mei Village was a newly built concrete road and easy to drive on. I was also quite surprised by the new concrete road that led from the entrance of Mei Village to the center, as it was broader and more durable than the one that I had seen in 2012. As in other villages, several new houses had been built near the entrance of Mei Village on former farmland. Villagers had also built more new houses within the village.

Meanwhile, I frequently encountered sand-selling sites beside the main road along the Maoling River. I also heard many villagers talking about river sand and about how some people extracted sand from the river near the village, which prompted me to visit the river, where I saw the dredgers. Also, I recognized that the river was deeper and wider compared to when I was a child. The changes in the river and in the mining of river sand therefore attracted my attention. Moreover, problems relating to the mining of sand persistently came up in my second visit, between October 2014 and March 2015.

First, some villagers, including the eldest daughter-in-law of my uncle, asked me to help them to stop illegal sand extraction from the river. The reason was that the land alongside the river was in danger of collapse, and some of the bamboo planted to protect the land was being lost due to the deepening of the riverbed. These villagers believed that I could ask government cadres to stop the illegal operators in the river since I had often visited some of the officials. The villagers told me to take a look along the river. When I went, I indeed observed the loss of land beside the river and the damage to the riverbank. Most households own land alongside the river, so the tension between the households and illegal miners was evident. Combating illegal sand mining was something they had already tried many times, and they had used every possible method, but had failed. The illegal operators continued to scoop out sand from under the people's land.

Second, several days after I had arrived in Mei Village, some villagers, and in particular some women, asked me if it was possible for me to help them to apply for grant projects or government money to build a bridge across another river near the village. The nearby river, which was once easily crossed by farmers who cultivated the fertile land on the other side, was now much wider and deeper than before due to illegal sand extraction. I was told that seven women had fallen into the river from their bamboo raft while heading to their field on the opposite bank. I actually did write several petitions and applications to the township government and to the Transportation Bureau of the district government, and also applied to the secretary of the CCP of the Qinzhou government through the online government system, but failed to elicit a positive response.

Third, whenever people talked about sand, they lowered their voices and spoke very carefully, and they always looked around and made sure that no one affiliated with the illegal sand miners heard what they said. Then, when my relationship with these villagers had deepened, they told me that in 2008 there had been a big fight within Mei Village and beyond, which was followed by further disputes because some miners were residents of Mei Village (Zhu 2018). As the days passed, I gradually realized that whenever two or more people spoke to one another, sand would always be part of the conversation. Even if I just came by for a chat, someone would still mention it. "A big piece of land collapsed last night; the stakes that I placed could not stop sand mining under my land," a woman complained to her neighbors and me during a morning chat. "I saw the dredger mining at about 6:30 this morning; they were extracting our land," an old man responded. "Half of our land disappeared, even entire bamboo trees were washed into the river," added another respondent (this conversation occurred in Mei Village in October 2014).

Last but not least, the ongoing mining of sand in the surrounding area that I observed showed me that the issue is a common one in many villages around the Maoling River. The most annoying thing for the villagers was that every day they could hear the sound of dredgers pumping sand out of the river while they were struggling with the consequences of the mining.

Thus, this period of field research indicated the abominable management of sand mining around the Maoling River. People told me that "the thieves" were stealing their sand. The ownership and use-rights of river sand and its relation to land, houses, roads, channels, dams, and the village were frequently discussed in the interviews. Diverse solutions by various agents have been attempted at the village level. Both villages and individual households have decided to sell their land because they could not stop sand theft on their land, on the one hand, while on the other hand the district (county) government or city government was conducting an auction to sell the river sand. Meanwhile, some local villagers decided to profit from the boom of a sand economy and either secretly sold river sand to dredgers at night, or bought dredgers to mine the sand themselves. Outsiders also came to extract sand, and while some had a license from the government or permission from the local village, others did not (Zhu 2018).

Others decided not to participate in or facilitate sand mining. One woman told me that her husband would never sell their parcel of land on the riverbank in Mei Village, even though it had already begun to collapse and was now smaller than before. She further suggested that by refusing to sell, her husband was protecting the temple of the land god and the trees around the temple.[6] If they sold their land, the overexploitation of sand would damage the adjacent temple of the land god as well as the sacred trees beside the temple. Among the Zhuang, the land god is considered to be the god who controls seasonal floods and protects the village's land and fields, as well as the god who guards people against evil. In her own way, the woman was pointing out what a good man her husband was to protect the public temple rather than exploiting the personal benefits (as illegal miners did) arising out of their land, because protecting the temple is a common value in these Zhuang villages.

Obviously, illegal sand extraction had not only influenced the relationship between the villagers and the village committee, as well as between the village and the township government, but had also influenced social networks, moral activities, and the belief system within the communities studied here. For these reasons, people kept talking about sand and, even though they lowered their voices, they still discussed the conflicts and possible solutions.

Furthermore, online investigations and an analysis of output in the local media (television broadcasts, newspaper articles, and local public websites), combined with further field research, showed that sand mining had become an issue of great importance in the nation. Yet, the increasing scarcity of river-sand impoverishment is visible by looking at the rapidly increasing price of sand, and the difficulty involved in mining and purchasing it, as revealed through my own survey. "They scooped out all the sand in the river," the informants claimed (interviewed in December 2014). The secretary of the CCP branch of Nagan Administrative Village responded to my questions about sand issues in February 2015, saying, "The river is empty; the natural village obtained money [from selling sand], but not the administrative village. Actually, we [the village committee] should also get some benefits from the selling of sand." He pointed out that discussing the issue of river sand did not make sense at this stage, since the sand was all gone. A wholesaler in Xintang Town pointed to the scarcity of river sand in the sand market in March 2017: "It is impossible to buy river sand since the price is so high and there is no river sand; apart from that, the governance is so strict." But during an online chat in January 2018, a sand miner of Huangwutun Town told me, "We will go to more remote towns where some parts of rivers still have sand."

---

6 The temple of the land god is a public temple built on collective land. It is located on the south side of Mei Village, beside the river. It is surrounded by privately owned land.

I originally planned to conduct a study titled "The De-agrarianization[7] of a Rural Village: A Case Study of Southwest China," but the villagers essentially persuaded me to study river-sand mining. No matter whether I wanted to hear about them or not, the issues of sand mining and sand management constantly appeared in my surveys during my fieldwork. Moreover, I had to somehow respond to the sand problems and to the requests the villagers directed at me concerning this issue. All of this prompted me to make river-sand mining the topic of my research. After taking this decision, I employed multiple methods to collect data on illegal sand mining in Southwest China. I elaborate on these methods in what follows.

4 Methods

After settling down in Mei Village, I began to conduct ethnographic research, engaging with the society by looking at people's daily interactions. I also used diverse field techniques—note-taking, audiovisual recording, interviews, examination of indigenous literature, and observation—which are the building blocks of participant observation (Falzon 2009). This is called a mixed-methods fieldwork strategy (Bernard 2006; Axinn and Pearce 2006), because several qualitative and quantitative strategies are used to gather field data and are triangulated in analysis.

The qualitative approach included participant observation, oral history interviews, storytelling, and focus group interviews. The quantitative approach was implemented by distributing fifty survey questionnaires in Mei Village in Xintang Town, and twenty-nine questionnaires in Naxia Village in Huangwutun Town. The fifty questionnaires in Mei Village mainly focused on local households' livelihood, farming, and wage-earning activities, along with education, house construction, and social interactions, in 2014. At a later stage in 2017, I designed a sand-related questionnaire exclusively for Naxia Village that contained three parts, and half of the questions required explanations rather than simple yes-or-no answers. Part 1 asked about the socioeconomic background of the informants, part 2 was concerned with the situation of private houses, and part 3 was about sand and its management. Respondents were selected by making use of disproportionate stratified sampling, the results of which I will present later on in this and subsequent chapters.

7 See Bryceson 1996.

Apart from focusing on several village areas, I used "following the sand" as a method for gathering broader data (see also Foster 2005). In addition, I conducted research in towns as well as in Qinzhou City and Nanning City to gather firsthand data concerning illegal transport, truck owners, sand dealers, sand markets, and sand-mining governance.

4.1 *Choosing the Key Informants*

Who is involved in sand-mining management? Who is affected by the decisions that are made? And who has been influenced by those decisions and by illegal sand mining? These are vital questions to ask when exploring the river-sand crisis. I used methods related to Actor-Network Theory when I investigated who was involved in illegal river-sand mining. Blaikie (1995, 207) argues that in the field of political ecology, observations about the actors' roles can make the transformation of socio-environmental systems more tangible and meaningful: "There is a political economic arena in which various people pursue their 'projects' with very unequal access to power in which to pack their own particular knowledge claim and to enroll others into their own project."

Actor-Network Theory has become crucial in the field of political ecology since the 1990s (Bailey and Bryant 2005; Lave 2015). It took me a rather long period to figure out who the key actors in sand management were in my research setting. People try to avoid being called "powerful," because it always carries the possibility of trouble, as power is never unidirectional (Giddens 1984). In the case of Qinzhou, this means that being accredited power can benefit certain actors, but can also change or harm them later on when disadvantages become overwhelming. Therefore, the relevant actors connected to sand-mining management are highly likely to hide their true ambitions and power. These actors are not only government officials, but also cadres of village committees or CCP branches, directors of village committees, heads of natural villages, production captains, elders, and the villagers who make complaints to the government, the villagers who were arrested by the police due to the violence, and villagers in general. Doing in-depth fieldwork and being persistent were the key means in eventually identifying the key actors.

For the purpose of analysis, I grouped the key actors into five categories based upon their relative hierarchic levels, listed here from low to high:

1) Natural village / production team: heads of the village, elders, and villagers.
2) Administrative village: village committee and party branch: director of the committee and the secretary of the party branch, other cadres.

3) Township government: mayors, deputy mayors, chairman of township people's congress, directors and officials of financial department and other agencies.
4) Deputy head of a district (county-level) government
5) Government officials of Qinzhou City

Conducting fieldwork in my village of origin was advantageous for my research, as it was easy for me to enter the field and build up many connections with diverse people. I could easily access villagers in Mei Village and other villages, and party cadres and also the members of the Nagan Administrative Village Committee. It nevertheless took me a long time to figure out who was involved in river-sand mining and selling. My rather long stays in the chosen villages and consistent participant observation allowed me to identify crucial actors and to map interlinkages between them. Thus, the long-term engagement that comes with participant observation proved necessary to gain access to this sensitive information.

In 2017, I learned through the Internet and TV news that there were conflicts over sand mining in Nachun Village in Naqiu Administrative Village of Huangwutun Town. In 2015, an illegal sand agreement had been signed between the delegates of the six production teams of Nachun Village and one local miner under the supervision of Mr. Nong, the party secretary of Naqiu Administrative Village. Most residents of Nachun Village were not informed about the agreement. Struggles took place inside the village. In order to bring justice to the village and capture wider attention, one villager uploaded a complaint letter to social media. I learned about this infamous conflict and intended to obtain more detailed data. I therefore went to Nachun Village of Huangwutun Town together with my sister, who worked at a TV station. She helped me to collect more data on cases of sand extraction, related conflicts, and their management. Also, she enabled me to interview some journalists and to follow cases of illegal sand extraction that were reported to the local TV stations.

My sister and I intended to visit the village of Nachun, but we mistakenly went to Naxia Village. We stopped and asked a man beside a road if he knew a village where there was a conflict over river-sand mining. Fortunately, this man said "Yes, I do. But where are you from and why do you need to know about the conflict?" I told him that I came from Xintang Town, which had similar disputes and conflicts. The man's response surprised me: "Nowadays, there are miners stealing our river sand as well. ... We are about to discuss this issue tonight in my village" (March 28, 2017). The man introduced himself as Mr. Hu, who turned out to be a local winemaker with a shop beside the road, and a respected elder of Naxia Village. With the help of Mr. Hu, I got access to Naxia Village and Nachun Village. We visited the sixth production team of Nachun Village, where I interviewed the villagers who were involved in the conflict and obtained original documents related to the illegal sand agreement. I also was shown the river section where illegal sand mining had taken place. With Mr. Hu's agreement, I was able to join the elders' meeting in Naxia Village in Huangwutun Town in March 2017, where they discussed how to sell their river sand.

During this meeting in Naxia Village, I learned that a neighboring village, Nazhong Village, was also selling river sand to miners. When I visited Nazhong Village, however, I had difficulty gaining the trust of some cadres and villagers there when attempting to discuss the sand-mining disputes. As Van Rooij (2006b, ix) wrote concerning his fieldwork on land and pollution regulation in China, "I know how difficult it is to accept a stranger in a community, especially when this stranger sets out to learn about local problems." With the help of a villager of Nazhong Village, I visited the home of the party secretary of Nadong Administrative Village, Mr. Huang, who helped me to enter the field in Nazhong Village. Mr. Huang was a key person in sand mining activities. We conducted in-depth interviews at his home as well as many telephone conversations during the mining period and afterward. He went to the river with me where we could see the mining activities permitted by Nazhong Village.

I tried to conduct interviews on my own with cadres of the Naqiu Village Committee, but they would not talk to me about sand mining. The secretary, Mr. Nong, had been involved in every important decision for decades, including selling the river sand to a private company, but the income from selling sand was only received by the village committee. The informants in Naxia Village and Nachun Village told me that Mr. Nong had always had a problem with corruption. The villagers embezzled money from the government's "Renovation of Dangerous Buildings" project, and poor households that applied for government funds to build a new house had to pay Mr. Nong a bribe ranging between 3,000 to 8,000 yuan, since the households needed him to pre-approve the fund application (based on the financial and domestic situations of the households in question). I visited Mr. Nong at his home to try to interview him. However, he did not want to tell me about river-sand mining, nor did he want to tell me anything about his personal involvement in the conflict. I left without obtaining further information. Later on, I went to Huangwutun Town with an officer in the city government, Mr. Qin, and with his help I arranged a focus group discussion with the Naqiu Village Committee.

Indeed, I believe that it is difficult in China to conduct field research at the governmental level without the help of officials. The higher the rank of the officers that a researcher can connect with, the easier it is to obtain assistance from lower-ranked officers. When I started my field research in October 2014, my supervisor wrote me an introduction letter that I brought with me to Qinzhou City. With the help of a relative, Mr. Liu, I became an intern at the Policy Research Department. When I submitted my supervisor's letter, they did not accept it, as they considered the introduction by Mr. Liu more important. The director of the department, Mr. Li, helped me to conduct my survey in an official way and to build relationships with related government departments. He had quite a lot of power and influence over lower-level officers, and he helped me to interview some other key informants and to gather data by sending notes to lower-level government employees and by accompanying me to Xintang Town in 2014. I then obtained detailed data from the Xintang Government.

That trip had a snowball effect on my ability to make connections. I became acquainted with Mr. Cai, who was the vice chairman of the people's commission of Xintang government and several government officials. In addition, Mr. Yang, who joined us on the trip to Xintang Town in 2014, was the deputy district head of Qinbei District government. In 2016, he was promoted to Qinnan District, where Huangwutun Town is located, and continued to help me with my fieldwork in that district.

I realized that obtaining direct information from the government of Huangwutun Town was also crucial for my research. I only gained access to the Huangwutun Town Government with the aid of an official of the prefectural government. A deputy mayor of Huangwutun Town helped me to gather data on illegal sand-mining sites and to build up a connection with local sand miners and sellers working along the river. Three other officials helped me to conduct a survey in Tuanxian Administrative Village and Nadong Administrative Village along the Maoling River. The cadres of Tuanxian Administrative Village and Nadong Administrative Village were crucial for this research. Most of these cadres were engaged in river-sand-mining themselves. The cadres were local villagers, but they had been selected by the township government or via an election as village leaders at the administrative level, and they received a salary each month from the government. Their actions will be discussed in detailed in chapters 6 and 7.

The illegal sand miners, the wholesalers, and the workers they hired were also crucial for this research. It was not difficult to get access to the local miners. I could simply visit the sites near the river where sand was sold and talk to the miners and workers there. The miners—insiders and outsiders alike—were very diverse, as were their sand-business chains. Some of them belonged to family businesses, in which all workers of the company at all stages, from sand mining to selling and transporting, were members of one family. They owned dredgers, trucks, and other machinery. Some of the sand dredgers came from other provinces and were owned by bosses of large sand businesses who had bought a sand license from the government for a long section of the river and had hired more dredgers to work for them. Talk of conflict between local sand miners and outsiders could be heard throughout my field research. I also interviewed some people who worked on the big dredgers, as well as truck and ship owners.

5 Multiple Positions: Conducting Research "at Home" and Beyond

A reflection on the multiple positions I assumed while conducting my research will help the reader to better understand my research outcomes. After I had settled in Mei Village for a second period of fieldwork in October 2014, many villagers were curious about the purpose of my research. They knew that I had studied abroad, but they did not know why I came back. One said: "You grew up here, and you know almost everything." I clarified that I had come back for my Ph.D. project—even though some of them did not understand what a Ph.D. project was—and that I was going to stay in the village for six months for the first phase. Generally, I did not mind being asked questions by my informants. At the beginning of my fieldwork research, however, they asked many questions about life and culture in Germany, rather than letting me ask my research questions. What did people eat there? Did they grow vegetables? What did the farmers in Germany look like? Were they really taller than we are? Do you speak their language? And so on. These interactions built up trust between my informants and me. In particular, by asking and answering, information and knowledge were shared between us. This reciprocity continued during participant observation. Fieldwork was not only a matter of taking but also of giving; later, when I became involved in helping the villagers with approaching the government, I was trying to give something back.

I was an insider when I conducted research in Mei Village, but an outsider when I did my fieldwork in other villages and cities. Bernard (2006, 347) draws the following distinction: "Participant observers can be insiders who observe and record some aspects of life around them (in which case, they're observing participants); or they can be outsiders who participate in some aspects of life around them and record what they can (in which case, they're participating observers)." Despite this differentiation, active participation with one's own body and mind allows both types of researchers to gain access to the meaningful psychological and physical experiences of others (Csordas 1988, 1993). The villagers and I shared feelings of frustration about the ongoing illegal sand mining and of disappointment with the failure of township and county officials to properly exercise governance over sand. Especially the physical experience of seeing the environmental degradation and the erosion of riverbanks made the fieldwork significant to me personally and showed me its relevance. I felt frustrated when the villagers said to me: "It is useless to do this questionnaire" (a villager in Naxia Village, March 30). Another villager bluntly said: "I can tell you all the things that unfairly happened to us. But it is useless. You cannot help us" (March 29, 2017). What a researcher wants is an informant sharing stories relevant to the research topic, but it is frustrating when informants question how a researcher can support their search for justice.

As a person of Zhuang nationality, I also am able to reflect upon the identity of the Zhuang ethnic group as well as the settlement patterns, customs, and ecological knowledge of the Zhuang villages in relation to river-sand mining. In getting access, it was an advantage that I speak the Xinli language,[8] Cantonese, and Mandarin, while I also understand the Zhuang language, which enabled me to freely interview people and gain access to diverse data. Moreover, my duty as a community member to protect common-pool resources, support public affairs, and to assist some poor households became important paths toward understanding the social dynamics of the village. Even though I conducted my fieldwork in my "own" village, this did not mean that it was an easy task to collect data or conduct interviews. Let me illustrate this with the example of Mrs. Pan, the wife of the former village head of Mei Village, who lived with her three grandchildren while her son worked in the city.

While I was engaged in a conversation with a former village official who had been in charge of land resources and labor management during the Mao era and the Deng Xiaoping epoch, it became clear to me that no one remembered the exact dates and methods of the land reform or the demographic situation in those days, since it was quite some time ago. However, Mrs. Pan's daughter-in-law told me that Mrs. Pan probably still had the original documents concerning the land reform. Therefore, from 2014 to 2015, I visited Mrs. Pan several times and asked for these documents, which allegedly were in the possession of her husband before he died. However, she told me that all the documents had been burned during the funeral of her husband simply because, according to local tradition, everything that belonged to the deceased was burned when they died.

8 The Xinli language is spoken by Zhuang people in addition to the Zhuang language and by other ethnic groups.

One day later, I called a doctor for Mrs. Pan when I happened to learn that she was lying in bed sick. She was touched by my help, because none of her three sons or their families had taken care of her (her sons were working outside the village at that time). To thank me, she showed me a great number of original documents that I had been seeking for some time. She shared with me a large bag of documents that included many different kinds of files from the People's Committee of the township, the Production Brigade (the Administrative Village) and the party branch of the CCP in the brigade. The files contained data on demography and land tenure during those periods.

I registered at the office of the Department of Policy Research in order to gain access to more government documents and to help the field research to continue smoothly. I was interested to learn that the department's officers needed to do a lot of short-term fieldwork and to investigate the city, counties, towns, villages, and harbors. Furthermore, my experience in the department provided me with a vast amount of information concerning policies. As a "member" of the department, I could easily interview some officers on the city level and below and expand my research areas and circle of informants. My connection with the government assured informants that I was trustworthy, and hence they gave me more internal information, even when my questions touched upon a sensitive topic. After trust was built, some officers became my friends, and they invited me to office banquets and to spend free time in the township areas. Significantly, this government connection gave me special power in the eyes of some villagers in Mei Village, who believed that I could help them deal with the sand crisis. Therefore, they asked me for diverse types of assistance and told me more stories about sand-mining issues.

The multiple identities I assumed during my fieldwork helped me to gain access to secondary data, such as government documents, and to political resources. With the help of my government colleagues, I could more easily interview the people who were involved in issues concerning sand, and I could even conduct focus groups with officials. Meanwhile, these identities also motivated the villagers to ask me for help concerning government actions in stopping illegal sand mining. By complying with this request, I became even more engaged in the sand-mining issue in Mei Village.

CHAPTER 3

The Commodity Chain of River Sand and the Ecological Consequences of River-Sand Mining

This chapter discusses the commodification of river sand in the GZAR. How did river sand become commodified, and how did it become gradually incorporated into a larger sand market? What is the role of "illegal" sand miners in the value chain of river sand? What are the consequences of the river-sand rush? In order to answer these questions, my analysis will take the growth of illegal mining into account, particularly considering that this illegal mining is now taking place on a large scale in various regions in China. The chapter then goes on to examine how the boom in the sand market has shaped the commodification of river sand and its commodity chain and has further formed sand-mining activities and management on a local scale. In line with the political ecology approach, the last part of the chapter calls attention to the severe ecological consequences of river-sand mining.

1 The Commodification of River Sand

Due to the construction boom, river sand has become a crucial raw material in China. This material is the foundation for various valuable and profitable goods, including real estate, public infrastructure, and private houses. The importance of sand is emphasized on the website of the Chinese Ministry of Industry and Information Technology:

> Sand for construction is a bulk building material product that consumes a lot of natural resources every year. It is a key raw material for constructing concrete skeletons. It plays a role in particle reinforcement and external pressure in concrete, which is related to the quality and safety of construction projects. In concrete structures, usually 6–7 tons of sand and gravel aggregate is needed for every ton of cement consumed. With the sustained and rapid development of infrastructure construction, China has become the world's largest producer and consumer of sand and gravel. In 2018, national sand and gravel production reached 20 billion tons, accounting for 50 percent of the world's total, with an output value

 | DOI:10.1163/9789004505919_004

> of nearly 2 trillion yuan, about 13,000 enterprises, and nearly 2 million employees.
>
> MIIT ET AL. 2019 (translated by the author)

This increased demand for river sand transformed its commodity chain. The labor and production processes of river sand in China are now highly commercialized. River sand extraction, transportation, wholesale distribution, and retail sale to final consumers are part of the daily market phenomenon, and they are highly transboundary in nature.

Rapid economic growth and attempts to attain rapid development are in various contexts worldwide associated with resource exploitation and environmental degradation. Political ecologists describe these processes with terms such as "land grabbing," "land rush," and "green grabbing" (Li 2014; Bumpus and Liverman 2008; Corson and MacDonald 2012; Sullivan 2013; White et al. 2012). When discussing "green grabbing" as a tool of government control, Li and Shapiro (2020, 95) argue:

> The rights of communities to riparian farmland and to the use of rivers for fishing and irrigation are often customary but unsupported by legal documents or land title. In China, despite post-Mao economic reforms, land still officially belongs to the state and is merely leased to households; it can be reclaimed at any time.

Many studies on resource exploitation and its consequences for the environment focus on land, but their analyses can also be applied to the extraction and use of other natural resources. The exploration of natural resources has become an accumulation strategy for some people; as Smith (2007, 26) argues, "Capitalism is more voracious than ever in vacuuming a supposedly external nature in search of commodifiable use values." In addition to capitalism, however, governments also play a crucial role in the commodification of natural resources. However, this is not a process that goes unquestioned. In the case of the river-sand mining industry in southwest China, the commodification of sand is consistently embedded in moral questions surrounding the extractive industry with regard to ecological as well as social and political justice.

I specifically make use of the concept of a "rush" to understand river-sand mining in southwest China. Sand mining in southwest China is characterized by initial large profits and an increasing competition between illegal miners, all leading to rampant and unsustainable extraction. As Duara (2015, 8) observed,

> The modern determination to conquer a disenchanted nature and subject the commons—natural resources such as water, air, and forests necessary to all life—and, indeed, all beings—to resource mobilization has also advanced in … Asian societies, leading to an unsustainable and dystopian future.

According to Li (2014, 595), "The characteristic feature of a rush is a sudden, hyped interest in a resource because of its newly enhanced value, and the spectacular riches it promises to investors who get into the business early. Hence the rush. Do it now before others spot the value, and the profit margins decrease." Li (594–597) explored the drivers behind the 2008 global land rush by discussing the role of "spectacles" and "discourses" well designed by private investment funds in order to attract investors. In comparison, there is no "spectacle" (596) deliberately crafted by investment managers behind the rush on river sand. Rather, the success of individual sand mines creates spectacles that attract others to the business. River sand is a commonly needed natural resource in China. The sand rush there is a by-product of the construction boom and rapid urbanization. This rush is a striking feature of China's economic development and natural resource management, as this chapter shows.

The sand rush reflects the rational and moral aspects of modernity, which is the dominant approach in southwest China, but this is not the only perspective on sand-mining. Various stakeholders have different moral economic perspectives on river-sand extraction. Sayer defines "moral economy" as "the study of how economic activities of all kinds are influenced and structured by moral dispositions and norms, and how in turn those norms may be compromised, overridden or reinforced by economic pressures" (Sayer 2004, 2, quoting Sayer 2000). The issue of the moral economy becomes critical in the river-sand mining industry in southeast China. I examine ideas about the morality and immorality of the "illegal" river-sand industry, one of the latest means to accumulate capital or become rich in the Zhuang region.

My findings on the increasing numbers of illegal miners and wholesalers show that the sand rush plays a key role in the commodification of sand in the Maoling River (Figure 5). The empirical analysis, from the village level to the city level, shows how river sand has been increasingly extracted in order to supply diverse public construction projects and private buildings. The commodity chain of river sand has become more mature at the regional, national, and international levels. In particular, the increased export and import of river sand reflects the construction dynamic and economic situation. I will discuss the economic and "rush" aspects first before discussing the ecological consequences. In what follows, I discuss the cost-benefit analysis within the commodity chain.

FIGURE 5 Two illegal sand-selling sites in Huangwutun Town
SOURCE: FIELD DATA, 2017

1.1 *The Increase in River Sand Prices and the Use of New Technology*

Due to the scarcity of river sand in the market, the price of one cubic meter of river sand in Xintang Town rose from 70 yuan in 2007 to 100 yuan in 2016, and finally to 160 yuan in 2019. Overextraction in the last decade has also contributed to the current scarcity of river sand and increasing sand prices.

Figure 6 is based on the testimony of Mr. Lu, who launched his sand business in Xintang Town in 2007. His sale site was the biggest and oldest in the township. According to Mr. Lu, between 2007 and 2018 most of his sand was imported from outside Xintang Township. The import price kept increasing over time, from 20 yuan per cubic meter to 42 yuan per cubic meter, and the transport fare also rose between 2010 and 2018. As more and more people joined the sand business between 2008 and 2015, the profit rates fell steeply. Moreover, the sand business was not consistent over the course of a year. The best time for selling sand was at the beginning and at the end of the year, when migrant workers came back with wages. They would buy construction materials to build new houses, or to add another floor to their existing houses. Therefore, Mr. Lu humbly claimed that he was able to earn around 30,000 yuan in 2016.

Furthermore, I was told by the sand miners in Huangwutun Town that the price of river sand had increased there since 2008 due to the entrance of big companies engaging in sand mining. These companies, originating from

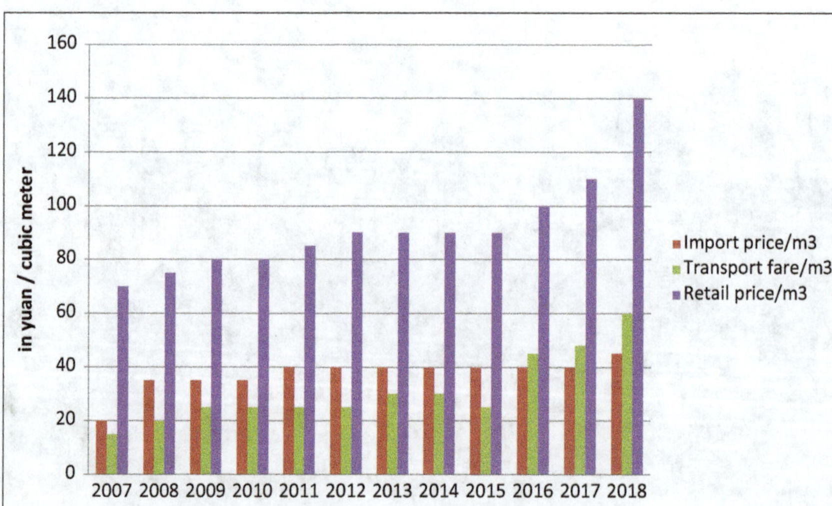

FIGURE 6 Business records of a sand importer in Xintang Town 2007–2018, showing costs of importing and transporting sand and retail prices
SOURCE: FIELD DATA, 2017

Zhejiang Province, Hubei Province, and Guangdong Province, bought river-sand mining licenses from the prefectural government through river-sand auctions (part of a kind of privatization of sand as a resource). These companies applied new technology and techniques, chiefly big dredgers, to effectively improve sand appropriation.[1] The emergence of new technology for harvesting river sand is a big change in the local area.

Sand extraction by large machines replaced the traditional harvesting models, in which sand is carried by hand or on the shoulders with bamboo baskets and shoulder poles. Sand-pump dredgers, sand pumps, sand-pumping motors, sand washers, sand conveyors, forklift trucks, dump trucks, and other machines are now commonly used when acquiring and transporting sand; these machines come in a range of sizes to suit different places, such as rivers, pools, seas, lakes, pits, and so forth, since there is a huge demand for dredgers and boats for sand mining. Many machine manufacturers have shifted their product lines to produce more sand-mining machines to suit the sand-mining industry, which aims to make more profits and reduce costs.

Downstream along the Maoling River, a dredger owner said, "Our new vessel with a 1,000-horsepower sand-mining pump can pump out 1,000 tons per hour" (Huangwutun Town, February 23, 2015). Another owner said that his 40-meter-long dredger could carry 450 cubic meters of sand, and that he could fill it in three hours. This technology helps miners to efficiently extract river sand. Such machines have made sand mining more profitable and thus have enabled the sand rush in southwest China.

2 The Sand Rush along the Maoling River

The sand rush along the Maoling River was driven by the long-standing practice of illegal mining. According to oral history, illegal sand extraction has been common in Huangwutun Town since the 1990s and in Xintang Town since the early 2000s.[2] Licensed mining only existed between 2008 and 2010

1 A small dredger can collect 20 to 25 cubic meters of sand each trip, while a medium-sized dredger can collect from 250 to 350 cubic meters of sand. A bigger dredger can collect 500 cubic meters of sand per trip, and the largest have a capacity of up to 2,000 cubic meters.

2 Sand mining started in Huangwutun Town earlier than in Xintang Town because sand from Huangwutun Town is exported to Qinzhou City to support construction there. Sand mining for households was open-access in Xintang Town, but only a few households had enough money to construct modern houses. Only after the construction boom started in 2008 were many rural residents in Xintang Town able to find jobs, and then they had money to build their own houses, leading to an increased demand for sand.

in Huangwutun Town and did not take place in Xintang Town until 2020. Dredgers of different sizes were mining sand, even during flood seasons, and the number of dredgers and dump trucks increased sharply between 2008 and 2016. The officials that I interviewed in the two towns confirmed that no one there held a government sand-mining license during my fieldwork period; thus the miners are considered illegal by the government. Another striking feature of the local sand rush is that illegal miners had to move fast, namely before the government enacted new laws and issued a restrictive policy on sand-mining management. Although new policies and enforcement actions against illegal sand mining have been strengthened since 2018, illegal sand appropriators remain the main suppliers in the local sand market.

2.1 *Illegal Sand Miners on the Upper Reaches of the Maoling River*

There were about ten sand-selling sites and sand miners along the Maoling River in Xintang Town in 2014, and also five dredgers working three kilometers upstream of the Xintang river section between 2013 and 2016 (see Table 4). All were smaller dredgers that could load only 20 to 23 cubic meters of sand at a time, which was suitable for the relatively small river section upstream along the Maoling River. The cost to buy these dredgers was between 20,000 yuan and 50,000 yuan, which was cheaper than the price of larger machinery in Huangwutun Town. The villages that were close to the river had one or two villagers who illegally mined sand for markets in various ways and during different periods.

Both dredgers and selling sites were strategically located beside the main road and the river, in order to transport river sand to consumers' construction sites. This location implies that illegal sand mining was an open secret in this region. As a result of the potential for economic benefit and the lack of licensed miners (see chapter 5 for the reasons why people do not apply for licenses), this enterprise has consistently attracted illegal operators. In the past two decades, the demand for sand has been filled with sand exploited illegally under the eyes of the village and township government. I was told that households that needed river sand for construction even ordered the sand in advance and were waiting for it. The river-sand harvesters could not fulfill the huge demand.

Since the 2000s, the river-sand business has become part of residents' daily life, and an important resource for some people's livelihood. These illegal sand operators have not paid any taxes to the government, nor have they paid any fee to the villages in the region. "Stealing" was the term used by local villagers and the local government to explain how the miners make a profit. Sand is in practice free of charge and open-access, which has led to free-riders marketing

TABLE 4 Illegal sand miners in Nagan Administrative Village

Name	Age	Occupation	Sales location	Dredgers	Other equipment	Employees	Sales targets
Mr. Fugeng	48	Farmer	Beside the main county road	1	One excavator and one bulldozer	Brothers	Local rural households
Pan Jiu	40	Farmer	Beside the village road	1	One excavator and one truck	Family members	Local rural households
Lei Wu	50	Farmer	Beside the main county road	1	One excavator and one truck	Family members	Local rural households
Zhu Xin	52	Farmer	Beside the village road	1	One excavator and one truck	Family members	Local rural households
Lu Chongba	43	Farmer	Beside the village road	1	One excavator and one truck	Friends	Local rural households

SOURCE: FIELD DATA, 2014

sand.[3] Furthermore, sand transport and marketing did not cost these operators any money, because others bought the sand directly from the wholesalers at the river and transported it in their own trucks or hired a truck owner to transport it.

The illegal sand quarries in Xintang Town were established in 2007 or 2008. However, many were closed by 2015, once the river sand was largely gone from Xintang Town. "There is no sand to mine in the river," stated the wife of a miner, Mr. Fugeng, in Mei Village. Mr. Fugeng shut down his sand business in August 2015, while Pan Jiu's sand quarry was still running in April 2017. Nevertheless, Pan Jiu had to shift his sand mining from common sand to private sand, which means that he bought private riverbank land in order to extract sand. For example, in 2016, he paid a Mei villager, Mr. Yaozu, 35 yuan per cubic meter of sand mined from his plot of land and resold it for 120 yuan per cubic meter.

Because of the overmining of sand near villages and towns, there is not enough river sand in the locality to supply all the households' building needs. Importing sand from towns downstream is the only option. The sand miners in downstream Huangwutun Town, which is located where the Maoling River enters the Maowei Sea, shifted their dredging places from rivers to the sea. Imported sea sand has come to replace locally mined river sand and is currently supplying the local sand market in Xintang Town.

2.2 *Illegal Sand Miners on the Lower Maoling River*

Huangwutun Town is one of the towns in Qinzhou City that is most affected by illegal sand-mining. At the peak of the sand rush, the town had ninety-one illegal sand-mining and sand-selling sites, as mentioned in a focus group interview I conducted on April 7, 2017. I interviewed the deputy secretary of the CCP, two deputy mayors (one of them female while the other participants were male), the director of the Construction Office (a man named Sange, who had been in charge of organizing illegal sand mining for more than fifteen years), and a prefectural official.[4]

According to the Huangwutun government's list of illegal sand miners, the number of miners had decreased to 41 by April of 2017. Moreover, local miners pointed out that there were 60 miners who had been forced by the government to stop mining by the end of 2018. However, the Huangwutun Township Government Working Report in March 2019 stated: "75 illegal sand mining

3 For the most part, mines sold the sand directly to local households, while others sold sand to marketing companies or brokers or only mined sand for a specific construction project.

4 The formal institution of sand mining at the township level was fully discussed in this focus group and is treated in more detail in chapter 6.

yards were demolished according to law and regulations in 2018." Variations in the reported number of illegal miners show the chaotic state of sand mining management.

Some of the dredgers in Huangwutun Town belonged to indigenous people, whereas the rest were owned by outsiders who came from Zhejiang Province, Hubei Province, Fujian Province, and Guangdong Province. This is different from Xintang Town, where only local people were involved in river-sand mining. These outsiders were called *lau lao* (外地人, those who do not speak the local languages, see section 5.1) by the local people.

Table 5 reproduces part of the government list of dredger owners in Huangwutun Town, which I received from a township official in 2016. All dredger owners were male and between 40 and 55 years of age. Each owned at least one dredger and had hired at least three local workers for sand-mining operations. The workers included several steady employees and truck drivers (the number of the truck drivers fluctuated, depending on the number of sand orders). Most dredging teams continued to extract sand the whole year round, except during the lunar New Year. This shows that the sand-mining industry is a high-speed industry. The miners worked hard in order to supply the market demand. Nevertheless, I observed in 2017 that some miners took a break during the Qingming festival. Mr. Huang, who was 55 years of age, said that since 1986 he had taken a daily break only from 9 a.m. to 4 p.m., and had harvested sand the rest of the time. His daily working hours are common for other miners too, as the dredgers can only operate during the high tide at night. Dredgers move along the river, as shown in Figure 7, a recent satellite image. The dredgers search for and follow the sand. But during the day, the dredgers are moored in the river while the staff is resting. It is necessary to mention that although the capacity of the largest dredger listed in table 5 is 250 cubic meters, some dredgers can hold more than 500 cubic meters.

The following case studies from Huangwutun Town demonstrate who is involved in the sand-mining industry and how sand companies have developed over time.

2.2.1 Case Study 1: The Family-Owned Sand Company

The 55-year-old Mr. Huang, who was one of the first sand miners in the area, indicated in an interview that there was no formal sand-mining governance in Huangwutun Town in the 1980s and 1990s. The original sand harvesters were local villagers who used simple technology to mine sand directly from the river in the late 1980s. At that time, they did not pay to mine sand and did not apply for any licenses. The sand was sold to customers in bigger cities where new buildings were being constructed. Due to the increased demand for

TABLE 5 A selection of dredger owners in Huangwutun Town

Family name	Age	Origin	Starting time	Break time	Number of dredgers	New or secondhand	Capacity of dredgers	Cost of dredgers (in yuan)	Hired workers
Zhong	52	Nanfu Village	February 2007	New year	1	New	230 m^3	1.63 million	4
Huang	41	Shengjiliao Village	April 2009	New year	1	New	——	1.63 million	4
Huang	45	Shengjiliao Village	May 2003	New year	1	New	230 m^3	1.65 million	4
Huang	40	Huangwutun Town	July 2013	New year	1	New	250 m^3	1.5 million	3
Wen	43	Tangxiling Town	March 2009	New year	1	Secondhand	160 m^3	1.35 million	3
Li	45	Huangwutun Town	June 2007	New year	1	Secondhand	180 m^3	0.7 million	3
Yuan	43	Qinbei District	2015	Daytime	1	New	206 tons	1.8 million	10
Huang	55	Huangwutun Town	1986	9 a.m. to 4 p.m.	1	New, with secondhand motor	230 m^3	1 million	6
Dang	52	Nanning City	2006	Daytime	2	One new and one secondhand	320 tons (new),180 tons (old)	6 million (new) 0.5 million(old)	8
Huang	50	Huangwutun Town	2010	Daytime	1	Secondhand	180 m^3	0.8 million	3

SOURCE: FIELD DATA, 2016

FIGURE 7 Dredgers near the mouth of the Maoling River in Qinzhou
SOURCE: GOOGLE MAPS, LAT.1.926314°, LON. 08.477118°, AUGUST 20, 2019

sand, the miners invested in new ships. They also purchased trucks and other machines in order to improve the mining output and transport capacity. At the same time, more laborers were needed for the production chain. Relatives and friends of the miners joined the sand-mining, transporting, and selling processes. In what follows I present the story of Mr. Huang and his family, who had been running a sand-mining and -selling business for 21 years at the time of the interview, to illustrate the transformation of river-sand mining through a historical lens.

Mr. Huang was part of the first generation of river-sand miners in Huangwutun Town. He owned a big sand-selling site, which was located one kilometer away from the township's major bridge. Mr. Huang began his river-sand mining in

1986. "It was the year my son was born that I began my sand mining and sand quarrying. Now my son is 21 years of age." At that time, Mr. Huang built his sand quarry with his wife on his own riverbank land. In the beginning, he only had a simple pump to directly scoop sand out from the river. Due to the rapid growth of the sand business and the benefits that he obtained, more laborers were needed. Several years later, his brothers joined him in sand mining. Mr. Huang's wife was in charge of account management. The family's sand enterprise was extended by purchasing a dredger, which was also brought to other locations along the river in order to harvest more sand. After about ten years, Mr. Huang and his brothers were able to build new houses and to buy new cars with their substantial profits from sand mining and selling. Their visible success attracted more people to the high-profit, low-cost sand-mining business. This is an example of the "spectacles" that Li (2014) referred to. When Mr. Huang's son graduated from a technical secondary school in Nanning City in 2014, he too joined the family business. Some outsiders also began to participate in the sand-mining industry. Mr. Huang always had ways to look for new, big construction projects and has never stopped his business: his successful enterprise helped him to collect enough money to buy a bigger dredger to mine ocean sand.

In 2016, he and his brothers went to Guangzhou City to buy a secondhand motor for this new dredger, which had more than 600 horsepower and was imported from Japan. Mr. Huang's son said that even though it was secondhand and expensive, costing 500,000 yuan, it also was more powerful than motors made in China. The family had to borrow money to pay for the new dredger and it would take three years to cover the costs. But their dredger was now also the most up-to-date of all the sand miners' dredgers in Huangwutun Town. The new dredger could load 450 cubic meters of sand each time, while the other dredgers could only load 260 to 350 cubic meters.

The family was very satisfied with the new dredger and Mr. Huang's brother gave me a tour (Figure 8). The ship is 40 meters long and has three floors. The third floor is a small control room containing a lighting system to illuminate the dredger at night. The second floor contains four large oil tanks to supply fuel for the dredger, a kitchen area, and a washing area (the workers go home after work). The first floor is a big storage chamber that holds the sand. Mr. Huang explained that each boat is equipped with one captain, one mate, two sailors, and one boatman. All members of the family are skilled workers who can deal with most situations that may arise on the big dredger. In terms of operating costs, Mr. Huang's son said:

> There are four large fuel tanks on the ship. Each sailing will use around three thousand yuan of fuel. The price of sea sand is 28 yuan per cubic

> meter. We earn around 10,000 yuan each time as net profit, and then they must cover the cost of labor, fuel, and so forth. Yet the business is not doing as well as before. The buyers do not pay us immediately. They pay after months or even longer. And sometimes, we can only earn 1,000 yuan each time.
>
> interview, April 10, 2017

Mr. Huang told me that they do not have a license for their business. The only required document is a sailing license from the Maritime Bureau, which inspects the ships annually in May. Not a single dredger or ship in the town has a sea-sand mining license. The long-term existence of unlicensed river-sand mining has benefited these miners, who believe that sand mining is a personal business issue, rather than a common, natural resource–related problem. "No one opposed our river-sand mining activities before. During these years of mining, we have not had big conflicts with others. Sometimes, small disputes took place, and a little money could settle them. The problem is the scarcity of river sand. That is why we bought the new dredger to mine ocean sand." Mr. Huang's son pointed out, without mentioning the villages or the township government, that when and how much to mine is dependent on the market and the family's storage capacity. Altogether, Mr. Huang's son was satisfied with his family sand business, saying:

> I prefer to work in our new ship, you know, it is my family business and I can gain a good salary from it. My wife and our son have also been taken good care of due to the well-run sand business. The women in the family do not need to work outside, but just take good care of domestic matters—for instance, the children. They do not need to do any agricultural work at all, since we can purchase grain. Nonetheless, my mother still insists on growing one-season rice by selecting good seeds of rice in order to eat heathier food.
>
> interview, April 10, 2017

2.2.2 Case Study 2: Changing Status from Worker to Owner

Yet another sand-mining Mr. Huang in Huangwutun Town had worked his way up the ladder. He was around 50 years old and had been running his sand-mining enterprise since 2010. At first, he had worked in a sand-mining company owned by someone from Zhejiang. Mr. Huang said that after the Zhejiang businessman had made a great deal of profit, the company sold its dredgers to him. The dredgers, which originally cost 2.5 million yuan, were sold to Mr. Huang for 800,000 yuan. He not only took over the dredgers, forklifts, and dump trucks,

FIGURE 8 The Huang family's new sand dredger
SOURCE: FIELD DATA, 2017

but also the clients of the Zhejiang businessman. He began to reap a profit after a year.

In 2016, Mr. Huang had one secondhand dredger, two forklifts, and six dump trucks that could hold 22 cubic meters each. He hired two workers, in addition to his son-in-law, who was also helping him. According to his son-in-law, six trucks were not enough for all the deliveries. Therefore, they had to hire some other trucks. In December 2016, for instance, Mr. Huang sent 30 trucks each night to a project and several mixing stations for cement, sand, and gravel in Nanning City, as well as retail to some Huangwutun sand-selling sites. Mr. Huang claimed that he had a sand license, but he failed to show it. He rented around 1,000 square meters of land located close to the river and to the city roads.

Mr. Huang's son-in-law helps to organize the deliveries of sand and to record the employees' working times. He explained that they did not pay taxes to the government, but they regularly paid the so-called "snake-head fee" (*shetou fei* 蛇头费) or "sand-head fee" (*shatou fei* 沙头费) to gain protection from a certain group or gang. "If you pay the fee, everything will be fine," he told me. "They can help you to construct all the relationships and eliminate all trouble."

There are four different groups that can demand a sand-head fee. First of all, there are local government officials from the prefectural to township level who require a "protection fee." This fee is gathered under the pretext of law enforcement, but the money will be distributed among a certain set of officials. Second, local villages collect the sand-head fee because the river-sand mining takes place in their territory—for instance, the village-owned beach, river section, and road (see chapter 6). Third, some local people, including drug users, gamblers, and people who have criminal records, have illegally and criminally collected money from miners and sand sellers by force. Finally, a fee is paid to the owners of land beside the river or the road who rent it out to the sand miners and sellers. The sand miners used diverse terms to label these groups; e.g., they used "gang" to refer to both the government officials and to the drug users, gamblers, and criminals. In what follows, I use "gang" to refer to the latter group only, even though all four groups collect their fees in an informal way and obtain this income either illegally or criminally.

The collection of sand-head fees shows the unequal power relations between the diverse stakeholders in accessing and controlling the river sand. The term "sand-head fee" was used by both villagers and government officials. Official government documents from Qinzhou stipulate that action should be taken against those charging this illegal fee, but the city has not yet succeeded in eliminating this practice, since the gangs have extensive networks, both inside the government and within the local communities.

2.3 *The Outsiders and the Insiders: The Boom in Illegal Sand Mining*

Sand mining in Huangwutun Town has become problematic, and part of this is due to the outsiders attracted by the sand rush. Concerning rushes in general, Horowitz (2015, 238) observed, "Once these resources are discovered by outsiders, their commodification may result in overharvesting for urban markets and concomitant inaccessibility for those who have long relied upon them." Nazhong Village's river section is about 700–1,000 meters long. This village has a relatively long history of sand "transfer stations" and sand extraction. Compared to the illegal sand extraction in Mei Village and Na Village, where extraction was performed by local villagers, i.e., the "insiders," the illegal sand operators around Nazhong Village are from other townships and even other cities and provinces, and are considered "outsiders." A field survey showed there were a total of 11 sand-selling sites around Nazhong Village (see Table 6 and Figures 9 and 10). These sand miners have formed a rather big sand market centered on Nazhong Village that draws truck owners from other cities in Guangxi Province. They employ both dredgers and sand extractors. Sand miners own dredgers and extractors and build an illegal site beside the river for storing and selling sand.

TABLE 6 Sand miners in the territory of Nazhong Village

	Name	Originally from	Length of operation	Dredgers	Extractors	Employees
1	Mr. Li	Guigang Town	6 years	3 (rented)	1	4
2	Mrs. Huang	Jianshan Town	7 years	1	1	Family members
3	Mr. Huang	Nadong Village	17 years	1	1	4
4	Mr. Lin	Dafanpo Town	4 years	1	1	3
5	Mr. Zeng	Guitai Town	8 years	2	1	7
6	Mr. Hu	Hubei Province	9 years	2	1	5
7	Hu Wu	Nearby village	10 years	1	1	2 and a family member
8	Mr. Zeng	Kangxiling Town	5 years	1	1	4
9	Mr. Zhang	Nanning City	4 years	1	1	Family members
10	Mr. Xing	Nearby village	8 years	1	1	2
11	Mr. Liu	Huangwutun Town	7 years	1	1	Friends

SOURCE: FIELD DATA, IN 2017

Ten of the eleven sites included in Table 6 were regional sand-selling businesses. Sand mined in Maoling River was sold at a site beside the river or transferred to other wholesale markets. Interestingly, Mr. Zhang was an employee of a construction company and was sent to Nazhong Village to mine sand for his company. He said that all the sand he mined directly supplied the needs of his construction company in Nanning City. This means that his sand would not be sold on the sand market or to any other company. It was cheaper for the construction company to mine its own sand than to purchase it in the sand market in Nanning City (interview on March 29, 2017).

FIGURE 9 Two big dredgers in Nazhong Village's river section (*a*) and an advertising sign reading "Supplying Three Sizes of River Sand" (*b*)
SOURCE: FIELD DATA, APRIL 12, 2017

FIGURE 10 Trucks at sand-selling sites
SOURCE: FIELD DATA, APRIL 12, 2017

2.3.1 Case Study of a Female Miner: No Worry about the Customers

Mrs. Huang was the only female miner in Nazhong Village. Mrs. Huang, who was originally from the village but was now married and lived in another town, had returned to mine river sand. Her background as a former villager helped her to enter the mining market, and she stayed with her parents during that time. Her parents also helped her to hire local workers. I observed that truck owners preferred to purchase sand from her. Thus, her sand business has been going well. Even though there were more sand-selling sites around the river and truck owners could profit from lower prices if they bought more trucks of sand at the same time, they still went to Mrs. Huang first. The secretary of Nadong Administrative Village told me, "Mrs. Huang makes a lot of profit. She has built two big houses with five stories in her husband's town."

I went to the sand-selling site of Mrs. Huang beside the bridge at Nazhong Village, where I found her talking to the owner of a big truck. This truck owner was from the Heng County of Nanning City and he was speaking in Mandarin with Mrs. Huang. Sometimes she answered in Mandarin and sometimes in Cantonese. My recollection of their conversation follows:

TRUCK OWNER: Hi, boss, I want to load the sand, can I drive my truck to your site?

MRS. HUANG: Your truck is too big to load.

TRUCK OWNER: What are you afraid of? It is not you who is going to load the sand but the forklift. I will pay you the money. I have five empty trucks today, if you agree with us to load, then I will tell others to drive here.

MRS. HUANG: No, I really do not want to load any sand into your truck. As I said, your truck is too big. Last time, it took a long time to fill up your truck. Most importantly, your truck has two parts and has a problem turning around, and in the end you will damage my road badly.

TRUCK OWNER: 1,900 yuan each truck, and I will buy five trucks' worth. Just clarify if you will sell or not? You want to do sand business but you do not want to load. That is incredible.

MRS. HUANG: I will load your truck only if you pay 100 yuan extra to load one truck, and that means in total 2,000 yuan for a truck. Otherwise, I will not sell any sand to you.

TRUCK OWNER: Incredible! What are you afraid of? What you need to do is to just help me with your forklift by pushing my truck a bit like last time, and then it will be fine. If you do not sell me at the same price, I will not buy any sand from you; I will not come back for sure.

MRS. HUANG: It does not matter. As I mentioned, your truck is just too long and too big to load and it does not drive back to the road as other trucks do.

April 5, 2017

The truck owner left with an empty truck. Instead of lowering the price, Mrs. Huang added 100 yuan for loading sand for each truck. Mrs. Huang really seemed fine with the loss of five trucks' worth of sand trade. And she gave me the same reason to explain why she rejected the truck owner. She told me that she does most of the work at the sand-selling site herself, including mining, so loading would eat up a lot of her time. But the real reason was that river sand was in high demand and she did not have any problem selling all of the sand she had available in a single day: the trucks kept coming and she could get her work done regardless. The truck owner went to other sand-selling sites in Nazhong Village that were more easily accessible.

2.3.2 Case Study of a Local Miner Restarting His Business: Performing a Rite before Mining

Mr. Huang's sand-selling site was the oldest (in operation for 17 years) among the 11 sites listed in Table 6. He originally came from Nadong Administrative Village, on the opposite side of the river, but had migrated to neighboring Fangchenggang City years ago. He started his business in 2005. However, his sand quarry was dismantled in 2008, when the Qinzhou Government sold that section of the Maoling River to a private corporation through the sand-license auction in 2008. Unlicensed or illegal miners were threatened with arrest by armed police if they did not stop mining river sand. Machines were used to tear down the sand quarries and to remove the dredgers. Mr. Huang lost his sand-selling site in Nazhong Village as a result. He did not continue mining in the same location but moved to other places beyond the auctioned-off section of the river due to frequent monitoring by both the licensed company and the government. Downstream along the Maoling River, many other sand-dredger owners started to work with the company that now owned the license. They still extracted sand but were now selling it in the name of the company, which would charge 7 to 10 yuan for each cubic meter of sand. This charge greatly increased the price of sand in the construction materials market.

A local sand miner in Huangwutun Town told me in April 2017:

> We local businessmen hate these big companies. They earn the majority of the profit from our sand. Their new dredger replaced our small dredger and also the method of loading sand. The production and transportation

> of sand has become mechanized. I knew that they earned a lot of money from the sand business, even though they also paid a great deal to the government.

He said that several, frequently violent, conflicts had occurred at that time between the licensed company and local unlicensed companies.

Mr. Huang reestablished his sand business in 2010 after the contract between the government and the licensed company ended. He first bought an old dredger and rebuilt his sand-selling site near the bridge in Nazhong Village. After he had made enough money, he bought another secondhand dredger for 700,000 yuan with two friends in March 2017. This dredger could load 260 cubic meters of sand at one time. He hired four villagers from his own village to work for him.

Like other local miners, the day before he first used the dredger, he asked a priest to choose a good day to begin and performed worship by burning incense, offering food and fruit, burning spirit money (*mingbi* 冥币) and lighting firecrackers, and collectively praying with all of the workers to the "god of ships" (*chuantou goog* 船头公). This god is considered to be responsible for looking after ships and other transport machinery, and is worshipped by those people who depend on dredgers, ships, and trucks (see Figure 11). The illegal miners can be seen as disregarding local norms and values, as they ignore the norms that allow other villagers to demand sand and reject their duty to protect the village's territory and common resources. Yet, their worshipping of the god of ships shows that they are afraid of earning blame from both local villagers and the traditional gods and thus try to protect themselves. Interestingly, the local miners would also take part in collective worship activities.

In 2017, Mr. Huang mainly extracted sand from the Maoling River and its tributaries—for instance, the Nadong River, a major branch of the Maoling River that is close to Nazhong Village and Naxia Village. He paid 8 yuan per cubic meter of river sand to the local villages; he did not pay for the ocean sand.

After the implementation of the River Chief System in Qinzhou City in 2017, when the management of river-sand mining became more rigorous, Mr. Huang and his two business partners moved upstream to another township section along the Maoling River. They bought a section of the river from a local village there. Mr. Huang explained his business expenses:

> We hired seven local people, four of them working in the dredger, one person for loading, two working at the selling site to organize the stuff. The salary is 4,500 yuan per person per month; we spent more than 40,000 yuan just to hire the workers. We, the bosses, do not get paid. But

FIGURE 11 Mr. Huang performing a rite for his secondhand dredger (*a*) and mining activities after worship (*b*)
SOURCE: MR. HUANG, NOVEMBER 2016

> ten of us work together each day. We also need to pay the rent for the site, 40,000 yuan for one year. We invested 120,000 yuan to construct the sand-selling site. Now we have signed the contract with the village for the three-year mining permit. We paid 900,000 yuan for permission from the village, 300,000 yuan each year. Now it is the third year of the contract. Nowadays, we buy river sand from the villages and the money gets distributed among the households and for public use.
>
> interview, August 2018

It is common for sand-selling sites or dredgers to hire two to four local workers, with family members and friends working together. The salary of a worker has been increasing over the last several years, and many workers also share a proportion of commissions when they have good skills in mining or driving. Even though the costs of operation have been increasing, there is still profit for the miners. "We will make enough profit if the government allows us to mine. My uncle is in charge of the construction of a highway, which demands a huge amount of river sand. We have supplied my uncle's road constructions."

In short, the relationship between outsiders and insiders in sand mining is a tricky one. In practice, outsiders do not care about the common property or ecological concerns regarding public resources, such as sand, bridges, dams, and riverbank land, when they are mining in the territory of villages. The insiders obey the customary right of the villages and landowners. Insiders intend to pay for mining and perform local rituals before mining. They also contribute to public worship events in order to obtain public forgiveness if they damage fellow villagers' land. For instance, the local miner, Mr. Huang, had to buy his sand from the villages. Many of the local miners have made deals with the local villagers for permission to mine. However, even when an insider mines public sand, it causes disputes and tensions inside the village, as I will discuss in chapters 6 and 7.

3 The Role of Sand Wholesalers and the Sand Market in the Commodity Chain

Sand wholesalers and sand markets are crucial parts of the sand commodity chain. They note customers' demands, depending on the various uses of sand, and transmit their needs to sand sellers and sand producers (sand owners, sand miners, ship owners, and truck owners). In this section I analyze the commodity chain of sand, in particular the role of sand wholesalers and wholesale markets in cities in Guangxi.

3.1 *A Sand Wholesaler in Qinzhou City*

Wholesalers play an important role in sand's value chain. They supply sand to all construction sites across the whole province. I conducted several in-depth interviews with a sand wholesaler between 2015 and 2017. Mr. Xie, who was 35 years old, originally came from Hepu County, bought his own flat in Qinzhou City, and started a business in raw building materials (gravel and cement). Mr. Xie began his sand business in 2008, two years after construction had started on eight sports venues in Qinzhou City in anticipation of the Guangxi Regional Games in 2012. At the time of fieldwork, he hired two drivers and owned two trucks, and sold sand to big construction sites and mixing stations. He explained the marketization of the sand:

> My sand is from various places, so there are different kinds of sand. I import freshwater [river] sand from inland areas and salty [sea] sand from coastal towns. The entire cost of sea sand is 50 yuan per cubic meter, including the cost of the sand and the transport fee. The import price of river sand is around 20 yuan to 30 yuan per cubic meter. The transport fee and sand price are almost the same. The sand business has several separate sections. Essentially, there are four or five major players and steps from the harvesting to final clients. Specifically, from the owner of the dredger to the riverside transfer stations, to the trucks, to the dealers, and then to the clients, which is the complete industrial chain. My business belongs to the third and fourth part of the total five steps of the commodification of sand.
>
> interview in Qinzhou's sand market, July 2, 2017

According to Mr. Xie, the profit from sand is decreasing. Several years ago, the profit was really high, even several times the current profit. With the increasing cost of hiring workers and the rising price of oil and other items, the dredger owners can only obtain a profit of around 10 percent, which is lower than the profits that fueled the rush on river sand. The retail and other sections of the chain are better off than the ship owners, as sand continues to be in high demand. However, Mr. Xie complained that the business was not doing well after several years of the rush on sand. On the one hand, more and more people were entering the sand business to share in the profits, making river sand increasingly scarce. On the other hand, the government at various levels was taking more actions to control illegal mining. "However, if I don't continue in the raw materials business, what should I do?" he asked. For Mr. Xie, the sand business is the best option for a successful livelihood.

3.2 *The Sand Market in Nanning City*

Nanning, the capital of Guanxi Province, has three big sand markets that have greatly supported rapid urbanization and the development of the local real estate market. I visited one of these markets, Sanyou, in 2017. This market had 17 sand companies that were busy importing sand from other cities to supply the boom in infrastructural construction. Sixteen sand companies were owned by businessmen from Hubei Province. Only one sand businessman was a local resident. These outside businessmen brought their families or relatives to Nanning to help them in the sand business.

One accountant from a sand company in Sanyou Market told me that the company could sell more than 800 cubic meters of sand each day between 2012 and 2014: "It is normal to sell 500–600 cubic meters." This had fallen to only around 300 to 400 cubic meters in 2017.

Interviews in Sanyou Market indicated there are two ways to transport sand to Nanning City: waterway transportation and overland. Both methods have been heavily used. The road toward the sand market was badly rutted by trucks "transporting sand around the clock," as it was described to me by three local peasants hired by the government to maintain the road and control the traffic.

I saw some trucks come to the sand market to load sand in the daytime, while more trucks with sand arrived at night. A hired worker in a sand company told me that this was because of traffic controls that only allowed transport during the night, before dawn. Figure 12 shows a truck and ship transporting sand to Sanyou Market during the day in March 2017. Figure 12 (a) shows several piles of sand of different sizes and origins. The original cement road on the site had broken down and there was a mud road in its place.

Figure 12(b) shows a boat in the river dispatching sand to Sanyou Market. Several other ships were waiting to unload sand. One worker, whose job was to convey sand from the ship in the river, told me that he would charge the sand company one yuan to convey one cubic meter of sand. "I load about 10,000 cubic meters of sand each month, but in some busy months, it would be about 30,000–40,000 cubic meters of sand. For instance, one boat can obtain about 500 cubic meters of sand and it will take me six hours to finish loading." He and his wife could reliably earn 12,500 yuan per month, but would earn much more in a busy month.

The price of sand in this market has been steadily increasing. In March 2017, wholesalers would sell each cubic meter of river sand as middle-size sand for between 85 yuan and 110 yuan, and sea sand for 120 yuan to project construction sites, while its import price was about 70 to 75 yuan. By the end of 2018,

FIGURE 12 Transporting sand by truck (*a*) and boat (*b*) at Sanyou market in Nanning City
SOURCE: FIELD DATA, 2017

these sand prices had doubled or even tripled, although the cost for mining and transporting river sand also increased.

The only local businessman in Sanyou Market told me that the market was facing demolition by the city government, to make space for city expansion and to reduce pollution and damage to the roads due to the overloading of trucks. However, these environmental problems seem insignificant compared to the pollution experienced in the rural areas that supply sand to the urban market. The river flowing through Nanning City, the Rong River, has been used for transporting river sand from other cities, such as Wuzhou City and Hengxian County, because sand mining in the Rong River has been forbidden in recent years. Therefore, as Figure 13 (*a*) reveals, the water in Rong River was clear and blue, even though some boats were loading sand there. In the background of this photograph are many new buildings that have been constructed to supply new homes for citizens and new migrants. In contrast, Figure 13 (*b*) was taken in Huangwutun Town, where the Maoling River was heavily mined to supply the demand for sand transported by trucks. Therefore, the color of the river there was yellow and muddy. Truck transport has caused damage to roads and bridges. The background of Figure 13 (*b*) also shows that many private houses have been built there, but they are small when compared to those in Nanning City. These two photos show the contrast between the rural areas where river sand is mined and the areas of government-planned urbanization where sand is in high demand. My interviews with sand wholesalers show that they supplied tons of sand, originating from Huangwutun Township and other rural areas, for the above-mentioned government projects.

3.3 *Truck and Ship Owners*

Transporters play a crucial role in sand's commodity chain. There is no single pattern for buying and selling. Miners extract sand from the rivers directly and unload river sand on the shore where the storage plots or initial wholesale markets are. From there, trucks and ships will transport the sand to diverse destinations, including urban and rural wholesale markets or construction sites.

For each trip carrying ca. 45 cubic meters of sand from Huangwutun Town to Nanning City, truck owners could charge 500 yuan. Owners of ships containing 560 cubic meters of sand could earn 10,000 yuan each time. However, an interview (March 28, 2017) with a ship owner revealed that the shipment was only profitable if he could carry sugar and other goods on the return trip. He said, "I have four workers. It takes us five days for one trip. The costs of hiring four staff and for the fuel are 20,000 yuan. In addition, I need 20,000 yuan to buy the sand. The sand can sell for 40,000 yuan."

FIGURE 13 Importing and exporting sand in Nanning City (*a*) and Huangwutun Town (*b*), 2017
SOURCE: FIELD DATA, 2017

In order to make a profit, trucks were often overloaded and owners and drivers disobeyed the traffic laws. One day I found Mr. Huang's dredger parked in the river beside his sand quarry in Nazhong Villager. There were trucks coming and going to transport sand. He pointed to a truck and told me:

> This truck can load 22 cubic meters each time. Before 2017, they used to increase the capacity of this style of dump truck by adding planks. By doing so, they could load another 22 cubic meters. Therefore, each time the truck can earn around 500 yuan. That meant the owner of the truck could earn more. However, this action caused the trucks to be too heavy, and they seriously damaged the bridges and road. The traffic policemen banned this retrofit. If a truck was found transporting with additional planks, the traffic policemen fined them 30,000 yuan.
>
> interview at MR. HUANG's sand-selling site in Huangwutun Town, April 1, 2017

The new, stricter sand-transporting traffic policy was published by Guangxi Government in March 2017. During field research, I heard many complaints about this stricter policy. When I interviewed some owners of the kind of truck Mr. Huang had described, they said: "We could not survive under the new policy, because the same truck uses the same amount of petrol but transports half the amount of sand. The government only knows how to collect fines and taxes."[5] Environmental protection measures in the region made the sand industry less profitable. Yet, before such measures were taken, the sand rush significantly contributed to the river's ecological degradation and the erosion of riverbanks.

4 The Ecological Consequences of Sand Commodification

Extensive commercial sand mining is a rather new phenomenon in southwest China, following rapid industrialization and modernization in the late twentieth and early twenty-first century. These processes have had severe ecological and environmental consequences regarding geomorphology, hydrodynamics, water environment, flood control, riverbed evolution and deformation,

5 The drivers did not say these words politely to me, but rather angrily. They believed that I was a journalist looking for a news story. "You reporters just show up for two hours to get the news story that you need, and return to your office. Meanwhile, we have to drive overnight and are afraid of the policemen. Otherwise, how can we feed our families?" (interview, Nazhong Village sand-selling sites, April 3, 2017).

bank erosion, and channel recovery in diverse rivers (Kondolf, Smeltzer, and Kimball 2002; Sreebha 2008; Jia et al. 2007; Han et al. 2005; Mao and Huang 2004; Zhang, Liu, and Huang 1996; Erskine 1988). The governing of resources remains a major problem, particularly in "developing" countries (Blaikie and Brookfield 1987). For example, Shapiro (2001) argues that the environmental destruction that took place in China under Mao was caused by the abuse of nature. Elizabeth Economy (2007, 38) states: "China has become a world leader in air and water pollution and land degradation and a top contributor to some of the world's most vexing global environmental problems." The role of the transformation of river sand's commodity chain has been an understudied topic in this regard.

In this part of the chapter, I focus particularly on the ecological effects of the over-appropriation of river sand, which is tied to the expanding market. I discuss faunal destruction (i.e., of fish species), water pollution, collapse of riverbanks, river degradation and consequent loss of farmlands, and damage to infrastructure that is heavily dependent on the water and riparian soils. The ecological consequences of sand harvesting are already increasing anxiety and confusion at the local level.

4.1 *Loss of Fish Species*

The most important effects of in-stream sand mining on aquatic habitats are riverbed degradation and sedimentation, which can have substantial negative effects on aquatic life. Meador and Layher (1998, 1) pointed out that in "fisheries, biologists often find themselves involved in the complex environmental and regulatory issues related to instream sand and gravel mining." Paukert et al. (2008, 624–631) found that "the effect of dredging on the distribution and abundance of fish has been mixed. … Therefore, agencies evaluating these effects have little information on which to base their decisions." However, their study on the Kansas River in the United States showed that "creation of low velocity habitats at dredged sites may not be suitable for large-river fishes, and efforts to reintroduce these fishes or provide suitable habitat could be hampered by these low velocity habitats" (631). For instance, the numbers of sport fish from the centrarchid family (e.g., bass, bluegills) "were reduced in dredge sites in the Brazos River, Texas and gravel-bed streams in Arkansas" (Forsage and Carter 1973, cited in Paukert et al. 2008, 624).

Fish are an important food resource for riverine people. For instance, my household survey in Mei Village revealed that at least 3 families out of 94 were heavily dependent on fishing for a living. A woman shared the story below, which resonated in most of my interviews:

> Apart from feeding my five children, I could also sell the fish that my husband obtained each day. Together with other women, we often went to Na Village and other villages to sell our fish, because our families could not consume all the catch. Primary school teachers constituted my main customers. However, since the death of my husband in 2016, I have not sold any fish. Before his death, my husband was already complaining about the increasing depth of the river following the excavation of sand from the riverbed for selling. He had to purchase new nets because the old ones were already too shallow for his fishing activities. The Hong Kouma fish and Pu Sha fish [locally translated as "fish lying on sand" and "fish playing with sand" respectively] are no longer found in this river although they were dominant some seven years ago before the expansion of commercial sand mining.
>
> interview in Mei Village's grocery store, April 10, 2017

Jia et al. (2007, 203) indicate the environmental consequences of sand mining in another Chinese region: "The riverbed level between Boluo and Dong'an was lowered by 3.95 m, 1m annually. ... With the lowering of river bed elevation, the water depth of the waterways also became deeper." There was general agreement during fieldwork that the two species of fish mentioned above prefer shallow waters. Sand overexploitation has deepened the riverbed, thereby disturbing the habitats of fish and interfering with their reproduction.

In addition, the Asian clam (*Corbicula fluminea*) survives in sandy and shallow-water conditions and is also a threatened species. Apart from its medicinal value, indigenous people claim the river clam has a high nutritional value. Oral history accounts from Xintang Township and Huangwutun Township indicate river clams were especially crucial food resources during periods of famine prior to the market-driven sand exploitation during much of the twenty-first century. Mr. Huang, a town official, qualifies the importance of river clams:

> We were catching and even selling river clams during tough periods in order to supplement our diet. There were times that my grandmother would be very harsh on me for harvesting too many clams. She thought that I was punishing her because she had to cook all of them and to find the right ingredients, including salt, for the meal. Some twenty or so years ago, clams were the food from the most celebrated rivers in our townships. They have fed generations, unlike dog meat, which is gaining popularity, and they are written about in books, thereby making them a special part of our culture. It is simply a shame that clams will soon disappear from our diet. Following the uncontrolled harvesting of sand,

> river clams are increasingly scarce. Local restaurants, which thrived on river clam dishes, have to gamble on their availability from local fishermen. In recent years, the price of clams has increased dramatically, from 2 yuan per 500 g in the 1990s to 25 yuan per 500 g in 2017.
>
> interview in Huangwutun Town, April 13, 2017

These price margins demonstrate the increasing scarcity of a once-coveted food across parts of China.

Due to its scarcity, fishermen are innovating technologies for harvesting clams. Zeng Pei, a 35-year-old man whose main employment is gathering clams, noted that the way people harvest clams has rapidly been changing. According to him, the new machines for harvesting clams have been popular since the 1990s, thereby replacing the traditional methods of harvesting with hands or hoes. The rise of this technology indicates the effect of excessive sand mining, which has changed the river environment. Zeng Pei noted that the sand extractors do not collect clams, only sand. However, the sand dredgers scoop up both sand and clams, and while some operators may try to separate the clams from the sand, they rarely do so because the numbers of clams have consistently diminished. As a result, they sell sand that may contain small numbers of river clams that could have bred future generations.

At the time of the interview, Zeng Pei had docked his boat for a couple of weeks following days of unsuccessful fishing due to the disappearance of the clams (see Figure 14). In anger he said:

> I have not driven the boat for more than ten days. Why should I keep doing it when I am clamming nothing from the river almost every day? On the other hand, I use up so much fuel each time with no returns … no sand, no clams.
>
> interview in Nachun Village, March 2017

These sentiments were echoed by other fishermen who had docked their boats out of frustration.

4.2 *Water Pollution*

Sand mining also pollutes the water within the rivers and in surrounding areas. In his online article "True Grit: The Sand Mining Debate," Stronberg (2016) notes:

> Sand mining operations are large consumers of water and mining operations can impact groundwater quality and flow. Concerns about the consequence of silica dust due to mining, storage, and transport of frac sands

FIGURE 14 Zeng Pei's boat for clamming
SOURCE: FIELD DATA, 2017

> have been raised. As mining operations occur exclusively in rural areas, the presence of a noisy industrial activity—often occurring 24/7—has been cited as having a negative impact on otherwise quiet communities.

In March 2017, I conducted 29 questionnaire-based interviews in Naxia Village. As they reported on the impact of sand extraction, almost all informants complained about the brownish color of the water. The majority of women claimed that they were not able to wash their clothes in the river anymore and that the river had almost turned into a death trap, as the depth increased following the excavation of sand from the riverbed. About ten people said that the water levels had dropped significantly, thereby affecting their irrigated farming activities. Others noted that over-harvesting of sand had even drained some of the water from behind the area's dams, which are considered to be the life-protection lines of the village (see below and chapter 5). A good number complained about the lack of clear institutional mechanisms to address these challenges, blaming their leaders at the village, township, and state levels for the predicament.

It was also noted that the dirty water has negatively affected the happiness and positive emotions of villagers. Nobody wants to live in an environment where huge trucks frequently transport sand, roads are wet the whole year, and dredgers make lots of noise scooping out riverbeds.

Indeed, sand mining impacts the villagers' ability to live well. There are diverse interpretations of the notion of "living well" in the research literature.[6] The notion of living well in this context is based on a memory of river health in the pre-extraction period, indicating that the current communities have been suffering from the degradation of their river environment. In contrast, sand miners or sand wholesalers might have a very different idea of "living well," namely making profits during the sand rush.

A retired primary school teacher in Nazhong Village told me that villagers and the school rely heavily on the water. He noted that other rivers had also been suffering, not only due to sand mining but also because of mineral mining upstream. Informants echoed his sentiments when they told me that many iron-mining sites, which had been built upstream along their river, were polluting the river water as well. Surface runoff washes iron into the rivers, thereby increasing levels of pollution.

However, the local government, which is keen to support iron mining for economic reasons, disagreed with the villagers regarding the best approach to limit the pollution from iron. Finding no solution from the local government, villagers were planning to report the matter to the city government.

Mr. Hu, a 70-year-old elder, accompanied me to the river to show me how polluted the water was. We were walking on the dam,[7] which was constructed to hold back water and floods (see Figure 15). Sand-selling sites were also built on such dams, while farmers were growing maize and rice behind them. Mr. Hu said:

> As you can see, the color of the water is yellowish and with a huge amount of dirt. Apart from the sand dredgers, commercial-based iron mining has led to serious damage to the river environment. School-going children and students have been warned to keep away from the river to avoid possible accidents and diseases. They can no longer swim in the river. When I was growing up, this river was a paradise for all children during the hot summer. As you already know, the sand is gone; sandy beaches have also

6 The interpretation suggested by Acosta perhaps fits best this situation of sand mining in riverfront Zhuang villages: "Living well ultimately has to do with another way of life, with a series of rights and social, economic and environmental safeguards. It is also reflected in the guiding principles of the economic system, characterized by promoting a harmonious relationship between human beings individually and collectively, as well as with nature" (Acosta 2008, quoted from Ulloa 2015, 326).

7 The dam of Naxia Village is the Dragon Vein.

> disappeared. The riverbed is one big hole now, the water is excessively polluted, yet dredgers and mining engines are still roaring.
>
> interview on the dam of Naxia Village, March 28, 2017

It is important to emphasize that although these changes have only occurred within the last three decades, the damage has exceeded anything in living memory. Even after the River Chief System was introduced, dredgers continued to mine sand in Naxia Village and the river water still had the same yellowish color, as my sister Ke Zhu observed in September 2018.

4.3 *Destruction of Riverbanks and Adjacent Farmlands*

Farming along the river is one of the leading ways of making a living in Mei Village. In this rural area, the farm plots that dot the riverbanks of Mei River are a source of life for many poor families. The peasants of Naxia Village ranked land and water as the most important resources for their livelihoods (see Figure 16). As already discussed, the land they refer to here is the riparian area of the river and its riverbanks. The popular foods grown here include vegetables, yams, bananas, maize, lychee trees, and herbs. The increasing sand mining has, however, led to the collapse of riverbanks, which then leads to erosion of the agricultural topsoil and leaves farmers with little or no land to farm. I observed that between 7 and 13 meters of the land surface (as measured from the river's edge) had collapsed into the river following massive sand harvesting in Tuanxian Administrative Village and Nadong Administrative Village in Huangwutun Town.

Traditional knowledge about agriculture and aquaculture is crucial for growing crops and for fish farming but does not fit the current environmental situation. The transformation of the river geography has negatively impacted the peasants' farming practices, which they have conducted for generations. The rapid shifting of the river geography lowers the water level, resulting in a change in agricultural water use and frequent flooding. The villagers remembered the last huge flood by pointing out which pieces of land had been submerged, and for how many days.

The collapse of riverbanks in the villages has been the leading cause of local-level collective action against illegal sand mining. Ironically, however, this damage has also been encouraged by poor farmers themselves, who have habitually sold the riverbanks to illegal sand-mining companies. Ownership of riverbanks is largely unofficial—riverbanks are handed down from generation to generation (some were redistributed during the land reform between the 1950s and 1980s). In almost all cases observed, individuals or families who farm on specific sections of the riverbanks have unofficially inherited this land,

FIGURE 15 The sand-selling sites on the dam (*a*), and big dredgers on the river at Naxia Village (*b*)
SOURCE: (A), FIELD DATA, MARCH 2017; (B), BY KE ZHU, SEPTEMBER 2018

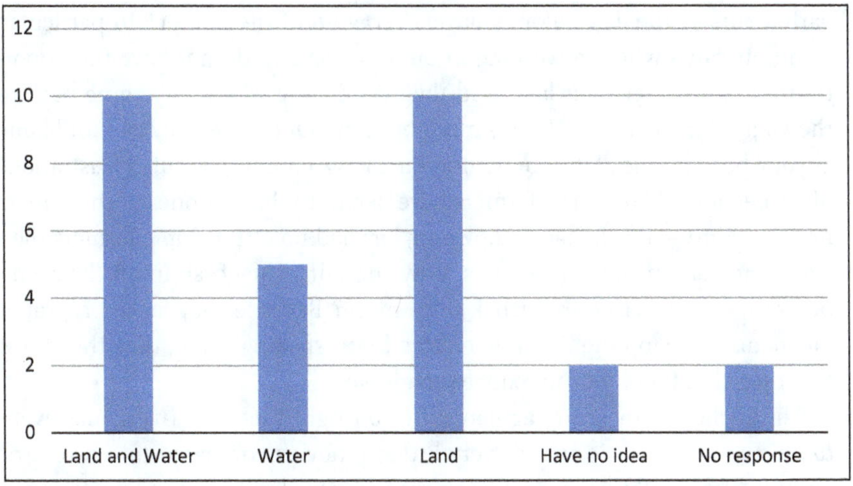

FIGURE 16 The most important natural resources, as ranked by residents of Naxia Village (N = 29)
SOURCE: FIELD DATA, 2017

officially owned by the natural village, from their ancestors. This form of ownership is culturally accepted and defended. Rarely would one farm a portion of land that was not his unless he had negotiated for access rights from the person on whom "ownership" of that land had been bestowed. Over time, such "landowners" have embraced the emerging market for riverbanks, which are believed to hold an untapped potential for sand.

These riverbank sales are not formal and the prices are not fixed—an illegal sand miner or a broker may approach a poor farmer who claims ownership of a specific portion of a river that the former hopes to exploit. At first, the owners may brush them off, claiming that illegal mining companies are destroying the riverbanks. However, such denial may be short-lived, particularly following peer pressure from others who have sold off their claims, or family members who are interested in making fast money. In most cases, however, the sale of riverbanks is quite secretive and may even happen under the cover of night. This situation often takes place when a seller does not want to be perceived by villagers as supporting the "dirty" business of sand mining. He may sell his portion of the riverbank one night, only to wake up the next morning complaining that some illegal sand miners had descended on his riverbanks at night. This is, in fact, a plausible scenario: Informants explained that illegal miners often descend on riverbanks under the cover of night and harvest as much sand as possible before daybreak. They then sell off the sand quickly and quietly in the early hours of the day, before villagers get wind of the matter.[8] In particular, some migrants who are working in urban areas and do not have the opportunity to look after their land find their sand gone when they come back to the village for the Lunar New Year celebrations. Of course, they cannot blame anyone because the illegal sand miners are long gone. The ensuing frustrations often become chaotic, and farmers have begun to throw stones at the miners in the vicinity—this behavior, however, has had no effect. Some farmers then started to sell off their riverbanks. Why shouldn't they cash in on their land before nocturnal "thieves" struck once more? Because they lacked a proper mechanism to stop the illegal operators from "stealing" sand, they thought it wiser to sell it for money to reduce their losses.

Thus, whether by secret agreement or outright theft, the riverbanks eventually change hands, often for very little money as compared to the returns obtained from selling the sand. After the payment has been used up, the poor farmers continue to condemn the illegal business without necessarily examining their own role in encouraging the vice.

In August 2015, an 80-year-old man, Mr. Yaozu, who opposed the illegal sand harvesting, eventually sold his riverbank land to a private company. During an interview, he complained:

> For eight years, I have tried my best to protect my land against those illegal operators. I have thrown stones at them to stop them from invading my land. I have made complaints to the local government, but they did not help me to punish those illegal operators. Nowadays, more and more people are selling their land to these companies; so do I. I am powerless to protect my land.
>
> interview, Mei Village, August 26, 2017

Meanwhile, as I discussed above and will explain in the next chapters, water-control levees, dams, canals, and irrigation systems have also been damaged by illegal sand extraction. The dam built in 1960 to protect fields and Nachun Village and Naxia Village from the river was damaged (see Figure 17). Flooding is a permanent concern, especially following heavy downpours. During interviews in various villages, the dam was referred to as the Dragon Vein (*long mai*

8 The demand for river sand is very high, and sand sells very fast in local villages. In the case of Mei Village, the buyers of river sand were local households, and they knew very well where the sand came from. River sand has become very rare in local areas, and each household wants to build its own house with river sand rather than ocean sand. Consequently, river sand is sold almost immediately after it has been extracted.

龙脉). The local villagers frequently highlighted the important role played by the Dragon Vein, which they considered to be a lifeline that protected the village and the field: "The flooding and the high tide will submerge our farmland and villages without the dam." They further stated, in the "Report on the Request to Investigate and Deal with the Village Party Secretary about the Illegal Sale of the Village's Flood Control Embankment":

> History has shown that Naqiu Administrative Village is the hardest-hit area and a high-incidence area of flooding. In 1998, the farmland of the whole village was flooded for ten days and nine nights. In 2001, the houses of more than ten households in the village collapsed due to floods. The state has invested tens of thousands of dollars to strengthen the village levee and to reconstruct water-damaged houses. Have the pain and the tragedy of history not given us enough lessons to learn? Do we want to stage this tragic history again?

Repeated flooding has shaped the villagers' behaviors and forced them to take action against illegal sand mining in their territory. The overextraction of river sand causes damage to the Dragon Vein, which has created conflicts between the village communities and the miners. Furthermore, as Pia (2019) has argued, memories of the history of cooperation in building dams and other collective irrigation projects in the Maoist era "leak" into the present and play a role as a motivating force that brings people together to find a solution for problems in communal resource governance.

Even though the villagers sold the river sand, they told the miners to keep at least 15 to 30 meters away from the dam, otherwise they would be fined. Of course, they did not believe that the sand miners would respect that rule—that is why they also went to the riverbank during the night after they had sold the sand to private miners. However, they did not succeed in stopping the miners from mining close to the dam.

4.4 *Loss of Life*

Riverbank erosion and subsequent floods do not only destroy riverbanks and riparian land but also cause fatalities. Peasants from Nachun Village showed me the parts of the dam where bodies had been found after a flood; the victim's houses had been located at lower elevations.

Deadly accidents also occur. River-sand mining often takes place out of desperation, and in dangerous conditions, since the miners work in every kind of weather, regardless of rain and flooding. An employee who was hired to work on one dredger died on the job. While dredging in the rain, he touched a live

FIGURE 17 The collapse of riverbanks in Nachun Village (*a*) and Naxia Village (*b*)
SOURCE: FIELD DATA, 2017

electric cable by accident. River-sand mining has also created a more dangerous river environment for local villagers and fishermen. In 2014, a woman, Mrs. Tang, fell into the Huangwutun River because a sand-mining ship crashed into a boat that she and her husband were tying up on the shore. Mrs. Tang fell into the river and the crew of the ship did not help her. Her body was found four days later. An accident less directly related to sand mining claimed the life of a boy who drowned in the stretch of river between Mei and Na Villages because of holes left in the riverbed by dredgers. One villager observed, "The river is not the same, as well as the river water. It is rather dangerous for our children, who are used to swimming in the river during the hot summer" (Mr. Hu in Naxia Village, March 29, 2017).

4.5 *Damage to Infrastructure: Roads and Bridges*

Apart from dams, river-sand mining also causes damage to other infrastructure. In Nadong Administrative Village, the now severely damaged Nadong Bridge once stood astride the river beside many big dredgers, but now villagers cannot cross the bridge anymore and must pay to cross the river by ferry. Note the brownish color of the river water in Figure 18.

The director of Nadong Administrative Village, Mr. Zhou, asked me, "Could you please help us deal with the bridge project? There are about 10,000 people who are impacted by this broken bridge, including the people of Fangchenggang City. We have been blamed for this problem by the villagers since 2015 when construction totally stopped."[9] I was told that the project ground to a halt due to a lack of financial support as well as mismanagement of the construction project.

Other impacts of illegal sand harvesting include the collapse of public transportation systems. Two bridges were uncrossable at the time of fieldwork. Two informants pointed to the danger faced by students who travel to school along the same roads used by big trucks and the same waterway used by mining ships. Some parents and grandparents used motorbikes to pick up students. "In summer, the road is very dusty, and on rainy days, the road is very muddy," a father who was picking up his daughter told me (interview, April 12, 2017, in Huangwutun Town). The increase in trucks has led to an increase in traffic accidents. The polluted river water and the noise of mining were also affecting people's health.

---

9 According to the Huangwutun Government Report issued in January 2020, "The Nadong Bridge with a total investment of more than 2.7 million yuan is to be resumed in April 2020. The bridge pier and bridge suspension have been completed so far. It is expected to be completed and commissioned during the Spring Festival."

FIGURE 18 The piers of the Nadong Bridge collapsed as a result of sand mining by dredgers (*a*). Villagers rely on ferries to cross the river (*b*).
SOURCE: FIELD DATA, 2017

FIGURE 19 Trucks driving through the wash pool to clean their tires of sand and mud before traveling on city streets
SOURCE: FIELD DATA, 2017

In Mei Village, roads and footbridges have disappeared following the illegal trade in sand. Ironically, in Huangwutun Township, many damaged roads and bridges are being rebuilt with funds from the government, which has not addressed the root cause of the problem.

In urban areas, the trucks ferrying sand end up muddying the roads. Consequently, Nanning City requires sand trucks to drive through shallow pools of water to clean their tires before travelling the main streets, but no other cities have adopted this policy. I counted eight trucks that drove through the shallow pool at the entrance of the Sanyou sand market within a ten-minute period at 8 p.m. in March 2017, indicating the large amount of trucks entering this city every day (see Figure 19).

In addition, the shift of dredging activity from river sand to ocean sand has also severely harmed the ocean environment, leading to the loss of sandy beaches, the disappearance of small islands, and the death of mangrove forests; it has also had a negative effect on the dolphins in Maowei Ocean in the territory of Qinzhou Region. Some of this environmental degradation is not visible within a short time frame, but the local people might have to suffer the consequences of these issues in the longer term. Furthermore, according to the Qinzhou Government's report in July 2018, Qinzhou will export ocean sand to support the construction of the new Three-Runway System (3RS) at Hong Kong International Airport. A description of the project states, "The construction entails reclamation of approximately 650 hectares of land, which will be completed in phases for the subsequent works such as the new runway, taxiway system, as well as a Third Runway Passenger Building with 57 parking positions" (Hong Kong International Airport 2016). The long-term impact of Qinzhou's ocean-sand extraction should continue to be explored in the future.

5 Conclusion

The commodification of river sand and the gradual incorporation of the sand-extraction industry into a larger sand market, from village level to township level to prefectural level and up to the national and even global level, have contributed to river sand becoming one of the most in-demand resources in the GZAR. This chapter's description of the transformation of the demand and supply of river sand has put the commodity chain of sand into a historical perspective and outlined the sand rush that began in 2008. Listing some of the consequences of a land rush, Li (2014, 600) writes: "There are many potential lines of fracture in the resource assemblages. … If promises of high returns do not materialise, investors might lose interest." However, he acknowledges, "The land would still be there, or more accurately, the ground would still be there." This is different with the sand rush, which is literally undermining the villages. The region-wide and nationwide sand rush has also fueled legal and moral debates in local contexts. Illegal miners violate laws by appropriating river sand in the villages without paying taxes or governmental fees and transporting it to construction sites in various cities and villages. Illegal sand mining also brought about social problems within communities. More violence and criminal activity occurred in local communities, including bribing officials, hiring gangs, and threatening villagers who opposed mining activities, thus putting pressure on public security. Consequently, moral tension and discussions on environmental justice (Holifield 2015) have frequently arisen around illegal sand mining (see chapter 7). Other social consequences are violent conflicts inside lineage-based villages and distrust of village cadres and other local government officials due to sand mismanagement.

The illegal sand miners have rather strong political, financial, and social networks that support their unlicensed mining businesses. Under a regime of ineffective river-sand governance, river sand has been openly seized to supply the local sand market in Qinzhou City for decades. Unlicensed river-sand mining has thus been going on for dozens of years in rural areas where local communities have had free access to natural resources for generations. The visible success of early miners is a "spectacle" that attracts both indigenous villagers and miners from outside. Both groups of illegal operators have had to deal with government officials on different levels, local elders, and the general public. I have explored the relationships between these diverse stakeholders within and beyond the commodity chain.

Some local miners performed religious rites with the intention of being able to operate their sand-mining business smoothly. This shows that they are obeying the customs of the Zhuang communities; under this logic, they also admit that the river sand is owned by the villages or individual households. That is why they have made sand-trading deals with the villages, in order to mine in the territory of those villages. Indeed, this kind of sand trade avoids the government's interference, and both the illegal sand miners and local villages have benefited from these arrangements. Nonetheless, there are those companies (government-owned, or owned by investors from other provinces) that can afford to participate in government auctions to obtain a mining license and then invest huge funds into better machines in order to mine the sand more effectively. On the one hand, the monopolization of river sand by bigger sand companies is a strategy for the commodification of river sand by the government and for proper sand governance. On the other hand, a member of a small-scale river-sand mining company would have great difficulty paying the high fees of hundreds of millions of yuan that each company has to pay, and thus small operators choose to mine illegally in the same river, which has caused even more tensions.

This chapter has also discussed the ecological consequences of river sand's commodification. Many villagers and also sand miners stated: "The river sand is gone." The scarcity of river sand has not only led to an increase in the price of river sand, but it also killed off fish species, polluted the water, and caused severe damage to the riverbank and dams. Crucially, the whole river's ecosystem has been reshaped due to the emptying of the riverbed, the collapse of the riverbank, seawater infusion, and other drastic changes to the geography of the river.

Significantly, the severe ecological consequences of river-sand mining are the same regardless of whether it is licensed or unlicensed. These consequences urgently need to be explored and investigated in great detail, since the degradation of the river ecosystem is gradual but ongoing, and may well continue to worsen over the foreseeable future. Nonetheless, the impact of sand mining is already all too visible: "the rivers die," "there are no fish," "the water is unclear"—all of these statements were heard repeatedly during my field research. Meanwhile, issues of damage to water infrastructure and related agricultural construction have also emerged in the rural areas. These disruptions have interfered with daily life and normal agricultural and fishery production activities in the affected communities, and even caused the loss of human life. What Pia (2019) calls the leakiness of historical collective memories plays a role in constructing public infrastructure in the present. Similarly, a historical communal memory of clear water and rich aquatic products in various rivers becomes a vital motivation for people to collaborate on finding solutions for environmental crises and resource misgovernance in rural areas.

Capitalism and the forces of modernity, including development and urbanization, have significantly impacted river-sand mining activities and exacerbated the environmental crisis in Southwest China. This situation reminds us of Ho's (2006, 6) question regarding whether it is possible to reconcile economic development and environmental protection. This research suggests that at least in the case of project-based development in Southwest China, a model that will be discussed more elaborately in the following chapter, environmental protection has suffered at the hands of economic development.

CHAPTER 4

Government Policies and the Commodification of River Sand

What causes changes in the demand for river sand? Why has this natural resource been appropriated and turned into a commodity in the construction market? This chapter seeks to provide answers to these questions by discussing how government policies, government-based projects, and economic dynamics have impacted the commodification of river sand and driven both legal and illegal mining operations. This chapter illustrates the reasons for the increased demand for sand by referring to government reports and other secondary documents. It also provides empirical case studies, based on data gathered through participant observation. The chapter explains that the increased demand for sand is largely an unintended side effect of economic policies of governments on all levels. The chapter starts with an overview of China's economic stimulus policies and their relation to the use of river sand.

1 Project-Based Development and Its Role in the Marketization of Sand

The provision of public infrastructure is a critical government activity (Wegrich, Kostka, and Hammerschmid 2017). Infrastructure is high on the agenda in many countries, for instance, Germany, Australia, the United Kingdom, the European Union, and China (Qu 2012; Davis and Walsh 2015; Kostka and Fiedler 2016; Gauke 2016; Williams 2016; Bridge, Özkaynakb, and Turhan 2018; Paterson-Jones 2019; Cen 2019). In Germany, a commission of experts was established to form a legally independent body for planning and building roads, a task that has evolved into a serious initiative of the federal government (Expertenkommission 2015). In the United Kingdom, the National Infrastructure Commission was recently established as an advisory body to build effective and efficient infrastructure (Gauke 2016). Infrastructure is closely related to nation-building, as shown, for instance, by the regulation and financing of transportation infrastructure in Spain from 1720 onward (Bel 2011). In China, the State Council's Ministry of Housing and Urban-Rural Development of the People's Republic of China has been responsible for construction administration since 2008. Moreover, as Wegrich et al. (2017,

 | DOI:10.1163/9789004505919_005

1) explain, "Today, infrastructure governance involves more than the provision of brick and mortar assets. It includes the development of complex infrastructure systems relying on advanced and sometimes untested technologies."

China's economic stimulus policy—project-based development—plays a crucial role in the development of infrastructure and in processes of urbanization (Huang et al. 2014; Y. Li 2016). In the last decades, project-based development has become a common means of development in China (Y. Li 2016).[1] Also outside China, the improvement of public infrastructure has been adopted as a key approach to achieving national and regional development goals. On the one hand, Hammerschmid and Wegrich (2016, 31) remark, "Without the financial, regulatory, and coordinative role of the state, infrastructure investment would not happen." On the other hand, the enforcement of these government-based projects has become crucial to achieve national goals. Michael Mann (1984, 185) speaks in this context of "infrastructural power." This is very true in the case of my research setting, where the power of such projects has become a strong force in the distribution of natural and social resources. More specifically, it is clear that the drivers behind the increasing demand for river sand are rapid urbanization, infrastructure projects, and housing development, processes that are driven not only by market forces but also by the state.

According to the "Report on the Work of the Government" of March 5, 2009, huge infrastructure projects aimed to tackle the local consequences of the international financial crisis and to promote stable and rapid economic development (Wen 2009). For instance, China's 12th Five-Year Plan (FYP) for the National Economic and Social Development of the PRC (FYP 2011–2015) projected that China's average annual GDP growth would reach 7% and that the urbanization rate was expected to increase by 4% to reach 51.5% (Kostka and Mol 2013). The 13th FYP (2016–2020), passed by the NPC in 2016, claimed that in 2015 the annual GDP growth was 7.8% and the urbanization rate 56.1% (K. Li 2016). Recently, Premier Li Keqiang (2020) stated, "For the first time, permanent urban residents exceeded 60 percent of the population." The level of

1 Examples of project-based development include the Rural Drinking Water Safety Project, which began in 2005; the Extending Radio and TV Broadcasting Coverage to Every Village Project, which began in 2004, and the South-to-North Water Diversion Project (see also Qu 2012; Zhou 2012; Y. Li 2016). On March 5, 2009, Premier Wen Jiabao pointed out that "[China] has increased investment in infrastructure and basic industries, and has built and started major projects in energy, transportation, and water conservancy. ... Meanwhile, the central government allocated 42.3 billion yuan to support the construction of ten key energy-saving projects and environmental protection facilities since 2008." Furthermore, the 13th FYP (2016–2020) launched a number of major projects to develop infrastructure, improve the distribution of industries, and achieve ecological and environmental conservation (K. Li 2016).

infrastructure has risen overall (Xinhua News Agency 2016). A. Wang (2013) describes the economic development as a hard target that is of the utmost importance in determining the state's overall development plan. In particular, in order to maintain economic development, in 2009 the government initiated large-scale investments and began to implement a two-year investment plan totaling up to 4 trillion yuan (Wen 2009). According to Premier Li Keqiang (2016), across-the-board improvements were made in infrastructure during the 12th FYP between 2011 and 2015:

> The length of railways in service reached 121,000 kilometers, of which more than 19,000 kilometers are high-speed rail lines—more than 60% of the world's total. The length of expressways open to traffic exceeded 120,000 kilometers, the eastern and middle routes of the South-to-North Water Diversion Project were put into operation, and the world's largest 4G mobile network was built in China.

Investment in infrastructure projects is also seen to be crucial to achieve the main targets and tasks for economic and social development over the 13th FYP (K. Li 2016). In terms of the tasks of the 13th FYP, Li (2016) said:

> We should also expand major infrastructure projects, with the aim of increasing the length of high-speed railways in service to 30,000 kilometers and linking more than 80% of big cities in China with high-speed railways, building or upgrading around 30,000 kilometers of expressways, and achieving full coverage of access to broadband networks in both urban and rural areas.

China's infrastructure projects have become a major economic stimulus. The so-called project system has been a crucial means to foster road projects, high-speed rail projects, unitization projects (such as building new college towns and shopping malls), and the modernization of rural public infrastructure, villages, and houses since the late 1990s. However, the always-increasing growth in annual GDP and the numerous projects led to a boom in construction in both rural and urban areas, causing a huge demand for raw materials. Hence, the exploitation of river sand has facilitated a project-based economy.

Qu (2012) argues that the project system has profoundly impacted governance in China because it aims to break through the constraints of the original bureaucratic system represented by the unit system and to contain the differentiation effect caused by the market system through such means as special financial transfers. On the other hand, Zhou Xueguang (2011) argues that although "project packaging" and other methods enhance the ability of local governments to solve practical problems, they may also lead to uneven resource distribution and weaken the foundation of the legal system and of social sharing concepts, which is not conducive to alleviating the contradiction between the authoritarian system and effective governance. Li Yaofeng (2016, 107) believes that a lack of response on the village level to current academic research on the national governance project system hampers the improvement and efficiency of the project system when projects are implemented in villages.

The government implemented project-based development, or the project system, by investing funds in specific target projects that would, for instance, stimulate the engagement of private enterprises, reduce the government debt, and create more jobs, all with the aim of maintaining growth in GDP. The Chinese government has aimed to make use of public–private partnerships (PPPs) in order to implement these public projects in what is termed "traditional infrastructure," "such as energy, transport, water conservation, environmental protection, agriculture, forestry and major municipal projects" (NDRC 2016). The adoption of this approach by the Chinese government does not come as a surprise, given that fundraising is a common problem for infrastructure development globally. The European Union too has turned to PPPs to make up for insufficient public funds and in the hope of making public services more sufficient (Barnier 2003). Such PPPs constitute a system for redistributing political and social resources, in particular taxes, in order to achieve government goals.

Huang et al. (2014) argue that the Chinese government, at all levels, focuses on speedy development and economic growth. The project system obeys the logic of combining political power and money and is based on keeping costs low and pursuing profitable projects. But this system has led to some detrimental side effects, such as the low quality of roads and irrigation canals. Huang et al. (2014) therefore conclude that the project system should match other institutions and establish a high-value orientation in order to have a modernizing and rationalizing effect.

Figure 20 shows my classification of types of projects within the project system based on their relevance to the river-sand value chain. It does not incorporate projects such as the construction of dams or army infrastructure, because these have not been implemented in the research region. Here I focus on the general types of projects that have commonly been implemented in my chosen settings. The first category consists of individual support projects in rural areas, including the improvement of housing and comprehensive farming subsidies. The second category is formed by public projects in rural areas, encompassing

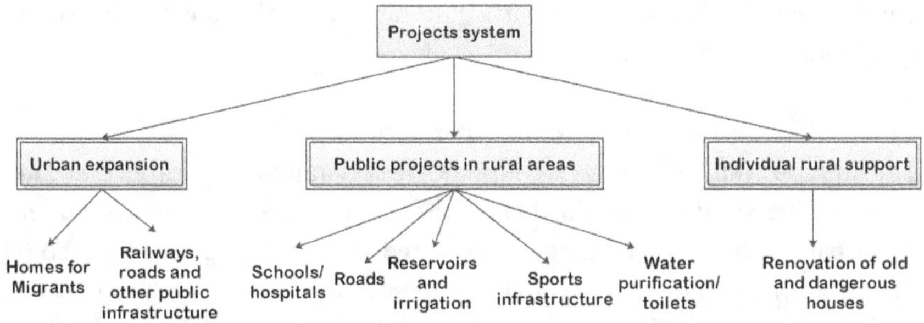

FIGURE 20 Overview of the project system
SOURCE: FIELD DATA, 2017; DIAGRAM: AUTHOR

the construction of roads, reservoirs, irrigation, sports infrastructure, schools and hospitals, water-purification projects, and public sanitation. The third category comprises projects meant to address the rapid urbanization that results as rural migrants settle down and buy apartments in urban and suburban areas. Among these projects, the construction of skyscrapers, shopping malls, subways/railways, roads, bridges, reservoirs, and irrigation canals require the most river sand.

Building on research on river sand in China (Xu 2011; Hou 2011), this chapter explores the impact of the government's development plans on the new status of river sand as a crucial and widely utilized raw resource in construction projects. Given this background, the rapid commodification of river sand has led to illegal sand mining in the research settings. The following section begins the analysis by discussing government-induced processes of urbanization.

2 Urbanization in Formerly Rural Areas

Urbanization is one of the major targets for achieving development in the context of China. According to Zhang (2013, 48), "The new urbanization is a grand strategy and historic task for our country's modernization and is the long-term driving force behind China's expanding domestic demand. It is also the 'locomotive' for promoting sustained and healthy economic development in our country." The rising rate of urbanization is driven by population growth, and it raises the problem of how to meet the needs of the increasing urban population with regard to infrastructure, accommodation, schools, hospitals, and food, among other aspects. Thus, the increase in urbanization shows how much governments at all levels have invested, which represents a direct and effective undertaking to expand domestic demand and provide employment (C. Lin 2014). In particular, rapid urbanization has consumed large volumes of river sand.

2.1 *The Incorporation of Rural Areas into Large Cities*

The urbanization of the GZAR has progressed rapidly over the last three decades. The New Urbanization Plan of the GZAR (2014–2020) pointed out that the urbanization rate in Guangxi is expected to reach 54 percent by 2020 (Q. Zhang 2015). Yu Zilong (2014) analyzed remote sensing data on urbanization in the GZAR. The study showed that the region's urban area expanded from 928 km^2 in 1998 to 5,120 km^2 in 2010. The area of land for urban use expanded by 1,244 km^2 between 1998 and 2004, and increased even faster, by 2,948 km^2, between 2004 and 2010. Among the 89 cities and counties, the capital city, Nanning City, had the largest urban expansion, reaching 885 km^2 in 2010 (Z. Yu 2014, 1).

In subsequent years, infrastructural projects, such as the construction of a two-line subway from 2013 to 2017 and the building of large transport hubs, rapidly increased in the province. In the capital city, Nanning, countless skyscrapers, streets, and subways are being built as part of diverse projects. Previously rural areas in the region have frequently been transformed into urban areas through such projects. For instance, the Nanning International Convention and Exhibition Center was built as the permanent venue for the China–ASEAN (Association of Southeast Asian Nations) Expo in 2004. The center has a gross floor area of approximately 130,000 m^2 and was designed in cooperation with a German company (GMP) and the Guangxi Institute of Architectural Design (GMP 2005). These buildings became a landmark of Nanning City and a symbol of urbanization, development, and international cooperation. The land for constructing these buildings was mainly appropriated by transferring collectively owned rural land to the state.

The incorporation of rural areas into the capital city of Guangxi is ongoing. A friend of mine, an official of Liangqing District in Nanning City, told me in September 2017 about another government-driven project stimulating urbanization, the 12th China International Garden Expo. Preparations for this exposition, which would open in Nanning City in December 2018, were frequently reported on by the media. The reports celebrated the success of Nanning City in having obtained the chance to host this national event. The government believed that it was a good chance to promote Nanning City's cosmopolitan status and to attract more investment to speed up urbanization. This Garden Expo was jointly organized by the Ministry of Housing and Urban-Rural Development and the GZAR. It was the first time that the Garden Expo was held in a minority region (Dong and Chen 2018; J. Liang 2018). According to a news report,

> The project is located on Dingzhao Mountain in Pumiao Town, Yongning District of Nanning City, with a total area of 658 hectares, including 263 hectares of the main park of the Garden Expo, about 15 hectares of the Dingzhao Mountain Protection Area, and 331 hectares of the idyllic scenery area. The garden has nine major functional zones, 80 exhibition parks of various types, and a total investment of about 3.957 billion yuan. It is planned to be fully completed by the end of November 2018.
>
> YANG, HUANG, and WEI 2016

It needs to be pointed out that most of the land used for the expo was previously rural and collectively owned. The land has been sanctioned for the construction of more roads, shopping malls, offices, administrative buildings, schools, and parks.

The dramatic expansion of the city has resulted in severe social problems, such as conflicts between the government and the dismantled villages. The construction of the abovementioned Nanning Garden Expo Park provides a case study of the consequences of city expansion in Nanning. I learned about this situation because many Mei villagers were involved in "construction for dismantlement." They had been hired as construction workers by the peasants whose land would be confiscated and whose houses would be demolished to establish the park. The tension between the local government and the peasants was high. A sand wholesaler from Nanning City sent me a video of the peasants in Pumiao Town who were hurt by armed forces during a mass demonstration against the confiscation and demolition of villages in the city's Yongning District. Migrant workers from Mei Village also told me about the situation:

> They [the government and the villagers] are fighting [for houses and land] every day; the villagers did not sign the compensation contract [prepared by the government] and the policemen were sent in to make sure that projects continued in a smooth fashion.
>
> interview in Mei Village, March 20, 2017

Under this construction-driven development, this area would be integrated into the capital city just as many other villages and towns were. The change in land use would also mean the status of the peasants would shift to that of city residents after losing their individual houses and their rights to use the collective land. This has created some issues for some rural households.

The compensation the rural households received from the government for the demolition of their houses was substantial but still not enough to buy a house in Nanning City, due to the rapid growth of the real estate market and exploding prices there. If a household accepted the fixed compensation, the house would be torn down and the ownership of the land upon which the residence had been build would be shifted to the government. The house owners had the option of resettling in new apartments (*anzhifang* 安置房) or special houses for a low price (about one-quarter of the price in the local real estate market) in other places of Yongning District, or receiving only financial compensation. An interview with two women whose houses were dismantled in 2015 revealed that the government had promised to build them new houses, but a year later they were still staying in a temporary apartment (*linshi anzhifang* 临时安置房).

If a household disagreed with the amount of compensation, or if the whole village considered the compensation offered to the villagers too low, disputes, conflicts, and confrontation would be the consequences. Some of these conflicts were filmed by local residents. Furthermore, videos uploaded on social media showed the police violently evicting people from their houses, hurting the residents in the process.

Ultimately, no one could defy the government and refuse to surrender their houses for demolition, but peasants found novel solutions to reduce their losses. "Construction for dismantlement" was the solution most frequently applied. It meant building a new house simply to demolish it again in order to receive compensation. The peasants gained more from compensation for a demolished house than they spent on building it. However, such solutions led to the waste of resources, such as river sand.

I was told by the workers of Mei Village that financial compensation was the major factor that prompted the peasants to engage in "construction for dismantling." The peasants were building houses in anticipation of profiting from the government compensation when those houses were demolished.

If a household possessed a title deed (House Property Certificate, *fangchanzheng* 房产证), the compensation for each household member, regardless of age or gender, would be 1,980 yuan per square meter. Without a title deed, it would be 1,320 yuan per square meter per person. Each individual would be compensated for a maximum of 80 square meters.

In Xinsheng Village of Pumiao Town, all households were hurrying to build more new houses so that they would obtain greater compensation following demolition. The village had more than 60 households, with the majority of these possessing more than two houses. While the government was busy tearing down the villagers' houses in order to support the "Twelfth China International Garden EXPO," the rural people hurried to hire more seasonal migrant workers to build houses to be demolished for compensation. Most of the houses in the villages were torn down by excavators hired by the district's government in 2017.

Seven villagers from Mei had spent the previous months working on building these impermanent houses. When I interviewed 24-year-old Mr. Jie, who worked with his brother and father in one of the villages that were eventually demolished, he said,

> There are so many households building their houses in this period, thus many are doing roof-cementing recently. We have to stay in the waiting line for the concrete mixer until next week.
>
> interview in Mei Village, March 20, 2017

Mr. Jie's father was the head of the construction team, which consisted of seven villagers from Mei Village who together worked on building two new houses. I asked them about the cost of constructing one of their working houses. The cost to build each square meter, including labor costs, was only 540 yuan. The cost of one floor with an area of 50 square meters was therefore 27,000 yuan. The huge potential benefits from compensation prompted the local peasants to build more houses. This case study illustrates the problematic impacts of project-based development on the natural and social environment of the research area—and in particular, the waste of construction material, including river sand and gravel. Furthermore, it also reveals where some migrant workers from Mei Village were working, and how their labor literally went to waste after only a few months.

2.2 *Rapidly Expanding Towns*

The development of townships is a major part of the urbanization policy, including in Huangwutun Town and Xintang Town. In this section I discuss government investments in public infrastructure and in public rental housing in these two towns.

Huangwutun Town has rapidly developed its township construction. The investments in what the government calls "social fixed assets" have been raised significantly, starting with an industrial park that was built in 2007.

Figure 21 shows the economic development of Huangwutun Town between 2007 and 2016. The data are not complete for all years, but it is clear that the township was expanding due to the constantly increasing investment, in correlation with the continuous increase of the town's GDP.

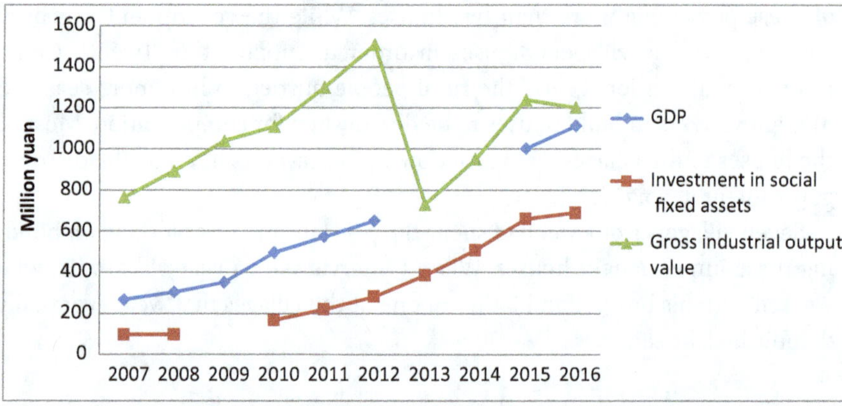

FIGURE 21 Huangwutun Town's GDP, social fixed assets, and gross industrial output value, 2007–2016
SOURCE: ANNUAL GOVERNMENT WORK REPORT OF HUANGWUTUN TOWN, 2007–2016

Xintang Township has likewise invested in social fixed assets, and I discuss its investments in some detail here. Xintang's built-up area covers 1.2 square kilometers, while the urban planning area covers a total of 2.8 square kilometers. The highway that connects Qinzhou City and Nanning City cuts through the town area. Since 2009, the government has improved three streets and a bus station and constructed three new residential neighborhoods. The township government contracted the Guangxi Haulan Design Institute to design and revise the planning of the town's development (QLCCC 2012, 434; AGWRXT 2013).

Between 2011 and 2013, the government made several other investments in the township. A second wave of development of real estate was started, and 80 percent of the new houses had already been bought by local rural residents by early 2012. The government also set up more than 50 spaces for parking on the street, the first official parking spaces in town. At the same time, the government raised 730,000 yuan to build public toilets and to improve the sewage and drainage system in the town (AGWRXT 2013, AGWRXT 2012).

The Xintang Township Government considered the development of such projects to be its main achievement. The "Working Report of Xintang Government over the year 2012" reveals that the township government had seven major construction projects, of which one was complete and five were under construction. Major projects had a total budget of 9 million yuan, of which 8.06 million yuan, or 89.6% of the budget, came from higher government levels. These projects were the construction of public rental housing for government workers; the construction of a training center for teaching planting techniques for *huaishan* (淮山, Chinese yams); the construction of office buildings for two village committees; a land-improvement project in three villages; the construction of two residential areas; and the construction of a so-called three-bird (chicken, duck, and goose) market in Xintang Town (AGWRXT 2013).

The "Xintang Township Government Work Report for 2017" shows that the township invested 430 million yuan in social fixed assets in 2016. The fixed-assets investment has increased rapidly compared to the investment of 285 million yuan in 2015. These investments caused a rapid transformation in the previously rural areas surrounding Xintang Township (AGWRXT 2017).

Apart from investing in public infrastructure, the Chinese government on various levels also makes investments in rental apartments in townships in order to calm down the overheating real estate market and to support the urban low-income residents:

> New public housing is not developed and controlled by employers or work units but by the local government in partnership with commercial property developers. ... The successful expansion of the urban housing stock through the market has been accompanied by a rapid increase in house prices, making home purchase increasingly unaffordable for low- and middle-income households and, in particular, for migrants and young office workers.
>
> CHEN ET AL. 2013, 14, 21

The construction of public rental houses (PRH, *gonggong zulin zhufang* 公共租赁住房) is one of the government-driven projects that have been implemented to improve the housing conditions for urban low- and moderate-income families. Public rental houses have a lower rent than the market rent for houses in similar conditions (X. Li 2011, 2). During the 12th FYP, the Chinese government aimed to build 36 million units of public housing (Chen et al. 2013, 14).

These projects were primarily subsidized by a vertical payment scheme, in which each level of government invests a proportional amount of money and the remainder of the cost is to be collected from the potential renters. These top-down payment transfers are an incentive to improve living conditions by "relieving the accommodation stress of local young staff and migrant workers who have relatively low income" (Li 2011, 6).

The town government, under the district government, has implemented similar projects to build apartments in townships and administrative areas.

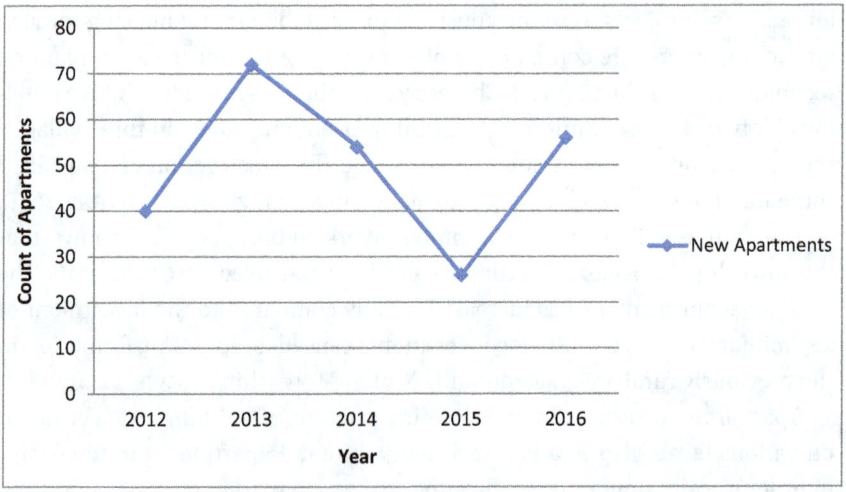

FIGURE 22 Public rental houses in Xintang Town, 2012–2016
SOURCE: ANNUAL GOVERNMENT WORK REPORT OF XINTANG TOWN, 2012–2016

Figure 22 shows the implementation of these projects in Xintang Town. I found that these public rental houses were in practice mainly supplied as apartments to public servants, including government officials, teachers, and doctors.

Together, all of these government investments in urbanization have driven the increased demand for river sand. In the following, I will present the construction of the administrative committee's offices in Qinzhou City as a further example of the large consumption of sand by the government.

2.3 *Rebuilding the Administrative Committee's Offices*

In order to enhance the development of the CCP and its service capability, the Party invested 510 million yuan directly from the central financial department in order to build new offices for its administrative villages and communities in Qinzhou City. I interviewed a city officer, Mr. Liu, about the office-building project called "One Workstation and One Center" (*yizhan yizhongxin jianshe* 一站一中心建设). He said:

> There are 1,010 administrative villages and communities, of which 986 have already built new offices. The goal of the construction was to accommodate party membership, resident services, social security, and social stability and so on.
>
> interview in Qinzhou City, December 12, 2016

On January 7, 2016, Mr. Zhang, who is an officer of the Qinzhou Prefectural Party Committee, told me, "The Qinzhou government created a new model of the workstation, which entirely promoted the grassroots Party construction on an advanced level."

The "One Workstation and One Center" project, which is promoted by the central government, includes the construction of physical infrastructure and the use of new communication technology—for example, the Internet and remote communication technology. By the end, the program had financed the construction of 244 Party Construction Workstations and 1,019 Village Party Service Centers in Qinzhou City. Construction was completed two years ahead of schedule.

The term "infrastructural power" was used by Michael Mann (1984) to describe the capacity of the state to intervene in and penetrate society in order to enforce policy throughout its territory. The successful "One Workstation and One Center" program strengthens the power of the CCP's branches. Indeed, the construction of village offices has significantly improved the efficiency of government propaganda and strengthened the implementation of the project system. Thus, the program not only increased the demand for sand through the construction of offices, but also through its support of the project system itself.

2.4 *Private House Construction in Rural Areas*

According to Xintang government statistics, by 2020, about 90 percent of peasant families in Xintang had built a new structure of steel and brick (see Figure 23a) instead of a traditional house made of timber, mud bricks, fired bricks and tiles, or lime (see Figure 24).

Table 7 shows that there were a total of 384 private houses built in 2013 in Xintang Township; 80 houses had been built in the urban community. Yet, in the villages too, the number of households residing in traditional housing was decreasing. For instance, in Mei Village, 14 new houses were underway in 2017, and the remaining households had built new homes by October 2018.

In my observation, the new houses have narrowed the quality-of-life gap between township and village households, despite the lack of employment opportunities in the villages. In fact, villagers are proud to claim that their living environments are much better than those of city and township dwellers, since the countryside has less noise and air pollution. Some rural residents do not have the capacity to build their own houses. Therefore, a national project called "Renovation of Dangerous Buildings" (*nongcun weifang gaizao* 农村危房改造) aims to help poor households build new houses if they meet certain requirements, such as a low income level or a sick household member. Statistics of Xintang Government from the year 2017 showed that 473 rural households

FIGURE 23 A new-style house in Xintang Town (*a*), and the gate (*b*) and front door (*c*) of another new house. Villagers have attached red paper on the gate and the door to secure blessings and protection from gods and ancestors.
SOURCES: FIELD DATA, 2017

FIGURE 24 Traditional housing
SOURCES: FIELD DATA, 2017

TABLE 7 Statistics on new houses completed in Xintang in 2013

Name of village or community	Number of completed houses	Area of completed houses (in square meters)	Value of completed houses (in million yuan)
Urban community	80	11,314	14.21
Nan ju	14	2,013	2.53
Tunlin	15	2,163	2.72
Ping kuang	32	4,549	5.71
Nagan	53	7,556	9.49
Tunlou	41	5,823	7.31
Tunwang	49	6,951	8.73
Nan zhong	57	8,095	10.17
Tuan zhong	43	6,138	7.71
Total	384	54,603	68.57

SOURCE: XINTANG TOWNSHIP GOVERNMENT STATISTICS, 2013

in Xintang benefited from this project between 2014 and 2016. In the GZAR as a whole, a total of 166,000 households obtained funds for dilapidated buildings in 2014, and the number increased to 210,000 households in 2015 (Tan 2015).

According to a document of the Qinbei District Government from the year 2015, a household could receive reconstruction subsidies granted by government agencies at various levels, as follows:

1) Central subsidy funds for rural reconstruction: average of 7,500 yuan per household.
2) Autonomous province matching funds: an average of 7,500 yuan per household.
3) City matching funds: 1,500 yuan per household.
4) District matching funds: 2,000 yuan per household.

Thus, an average household could obtain a maximum of 18,500 yuan in total from the different governments. Households themselves also needed to raise money to build their own houses. If an application was confirmed, an officer visited and inspected the house, and photos were taken before it was dismantled; the local government paid 9,600 yuan to begin with, and the other half was only paid after the new house had been completed and had passed inspection. Photos of the new house were taken as evidence.

Mr. Rui is a member of one of the poor households who obtained house-reconstruction funding in Mei Village in 2014. He married a mentally disabled woman and they had two sons and two daughters. All of them have a slight mental disability. The daughters got married at a young age, when they were around 16 years old. However, his two sons were still single in 2016. His elder son was 34 years old in 2014, but they could not find an appropriate woman for him to marry. Because of their children's mental problems, they could not find jobs for their sons at building construction sites, where other villagers work. No villager would want to work with them in construction jobs because they do not know how to build and cannot do the heavy work required on construction sites. Local officials told me that the family was always supported by the local government. They even applied to the government for rice and oil at the end of the year, when some families would have surplus rice to sell in the rice market. Even though they had no money, they still wanted to build their own new house because they hoped that their sons could get married in the near future, as others' sons had. "Any household needs / is able to build a new house" is a slogan used by the local government to support the goal of socialist rural construction, and so this family too was supported. Mr. Rui told me that he began to carry sand for his new house several years ago. With the government's support, by carrying his own river sand, and by selling off a large piece of inherited land to another household for 40,000 yuan, which Mr. Rui shared with a brother, each receiving half of the money, he eventually succeeded in building his house in 2014. This example shows that governmental policies also enabled the private construction of housing in rural areas.

3 Infrastructural Projects in Rural Areas

As mentioned above, the project system was not only executed in urban and urbanizing areas but has also created investments in rural public infrastructure. Several types of infrastructure projects, such as the construction of water reservoirs and road improvements, have also required sand.

3.1 *Water Conservation*

From 1949 onward, the government of Xintang Town built several reservoirs, irrigation channels, and other agricultural infrastructure to improve the production of paddy rice and other crops. Most of these were constructed between the 1950s and 1960s using free labor from local villagers (Li and Pang 2000). After decades of use, those facilities had become damaged. The central government decided to repair old facilities nationwide and has invested a great deal of money in agricultural infrastructure in general. In 2008, China invested about 160.41 billion yuan in water conservation, which was an increase of 56.3 percent compared to the year before (Wen 2009).

In 2008, the Xintang Government invested 8 million yuan in improving eight water reservoirs, primarily financed by the higher-level government, with the remaining part of the project funds being supplied by the Xintang Government. Two reservoirs were completed, reinforced, and repaired by 2015, because the Xintang Government was among the Fifth Batch of Central Finance and Small-scale Irrigation and Water Conservation Key Counties of Qinbei District. This also included support for the construction of an irrigation channel connected to the Fenghuang Reservoir (the reservoir mainly for Nagan Administrative Village). The construction scope included six administrative villages; the total funds were 20.35 million yuan. The project would improve irrigation canals of paddy fields to supply water to 16,300 *mu* (10.87 km^2; the total area of paddy fields in Xintang Township is 17,100 *mu*, about 11.4 km^2) (AGWRXT 2010–2017).

Meanwhile, water purification projects have also made progress. Three village committees completed their project aim of providing clean drinking water in 2011, while clean drinking-water projects in two other villages were under construction.

3.2 *Two Case Studies of the "One Project, One Discussion System"*

The "One Project, One Discussion System" (*yishi yiyi zhi*—事—议制) was originally created in the Chaohu area in Anhui Province in the 2000s. The rural residents of Chaohu invented this approach to raise funds and recruit labor to develop their village. Specifically, the villagers discussed one project at a time with a democratic approach, and decisions had to be approved by half of the villagers older than 18 or by two-thirds of all households. The provincial government later adopted this approach and implemented it throughout the province. Subsequently, the Ministry of Agriculture also adopted the approach as a new method to develop rural areas. The State Council issued the "Notice of the Ministry of Agriculture on Measures of the Administration to Raise Funds among Villagers" (GOSCPRC 2007). The notice defined the scope and number of participants required and suggested the procedure and management method. This "One Project, One Discussion" system has also been implemented in the GZAR since then.

According to the mechanism of the "One Project, One Discussion" system, the village is responsible for collecting the contributions required by the government's project system. That is, 75 percent of the funds of a project are provided by the government, while 25 percent of the projects should be paid for by the village in question. Moreover, all the land needed for implementing a project must be made available by the villagers. The system works according to the principle that those who benefit should contribute. Villages' contributions include not only money but also free labor, land, and organizational resources that are used to execute the rural development projects. "One Project, One Discussion" seems to be a cooperative system for building infrastructure in rural communities. However, the reality of infrastructure governance and project implementation is more complex.

In Huangwutun Town, road construction projects have been consistently implemented for several years. The investment in roads has recently been decreasing, as most of the roads had been paved with cement by 2012 (see Figure 25). Nonetheless, roads and bridges have been adversely affected by sand transportation in the township. Tens of thousands of trucks have entered the town to load and transport sand day and night. I frequently encountered traffic jams on the way to my research villages, particularly due to damaged bridges and the fleets of large trucks. Rural residents told me how horrible it was to face these circumstances day after day and even year after year. Their public roads and other infrastructure are damaged by sand mining and transportation. Even the governments have paid attention to the severe damage caused by trucks and dredgers. At the same time, road construction has continued, and more roads and bridges are being built, even though their maintenance is problematic.

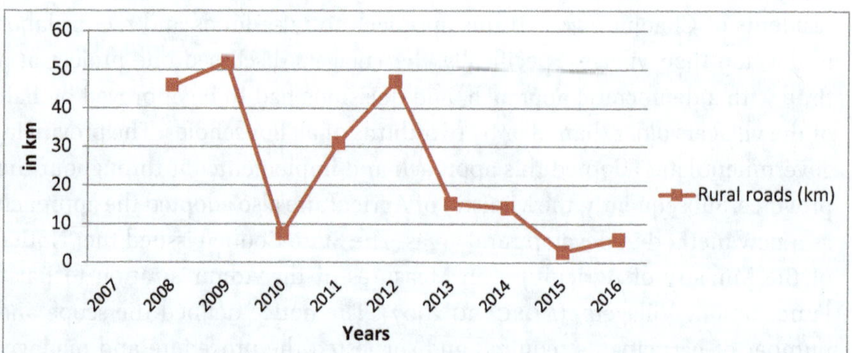

FIGURE 25 Road construction in Huangwutun Town, 2008–2016
SOURCE: ANNUAL GOVERNMENT WORK REPORT OF HUANGWUTUN TOWN, 2008–2016

The government of Xintang Township had completed 35 kilometers of paved village road (fourth-level highway) and 380 kilometers of tractor road by 2010, and thereby locally achieved the project goal of "extending cement roads to each village" (*cuncuntong* 村村通). This is a national project system, launched in the early 2000s, with the aim of linking all administrative villages to towns. The rest of the roads were also to be upgraded—for example, the rural roads in the villages.

According to the Annual Work Report of Xintang Government, 22.6 kilometers of cement roads in the 13 natural villages were completed in 2012. This cost approximately 280,000 yuan per kilometer of cement road. The Xintang Government thus invested 7.5 million yuan in social fixed assets, i.e., road projects. All of these rural road projects used the "One Project, One Discussion" system and were thus partly financed by villagers. According to Wegrich et al. (2017, 2),

> The governance of infrastructure—the planning, financing, contracting, and building of the public physical infrastructure essential for economic and social activities—is facing a somewhat paradoxical situation: while governments have become more reliant on private agents, their role remains critical. … And while governments may have become less visible in infrastructure provision, the job of infrastructure governance has grown more complex and ambitious.

Indeed, the government at all levels plays a big role in the project system. However, the role of village heads and rural individuals has been highly neglected. Two case studies set in Mei Village will illustrate the circumstances surrounding infrastructure governance and implementation.

3.2.1 Case Study 1: A 390-meter Cement Road Project in Mei Village in 2012

The elected village head, an unpaid volunteer,[2] plays an important role in the project system as well as in Mei Village. He may represent the village as a whole when cooperating with the government and dealing with diverse issues pertaining to projects. The major part of his job (the village head is

2 There are three villages in Nagan Administrative Village without a village head. Few villagers like to take this unpaid job. Two village heads claimed in interviews that they would like to quit their positions, including the village head of Mei Village.

always a man) is to make sure that there is land available for road building and public infrastructure construction. The new village head in Mei Village in 2012 was the person responsible for organizing the villagers to sell their river sand as a common resource to a private sand company in order to obtain public money. This collective income would qualify as the village's contribution of 25 percent of the cost of a road project within the "One Project, One Discussion" system. With the permission of the mayor[3] of Xintang Town, Mei Village was chosen as the recipient of a cement road project totaling 390 meters in length.

The village head stated that he made a great effort to convince households to contribute their land for the road and for other public infrastructure, but in vain. People wanted to benefit from the state's compensation for the occupation of their land. However, the government did not have funds to spare for compensation, even though the village was providing land for public use. Although the conflict was not well resolved, construction still went ahead and the roads were finished. In the end, the construction project was completed successfully—at least from the government's perspective, although not all villagers were of the same opinion.

After the new road was successfully paved in Mei Village, the government considered that village a good place to implement new policy. They wanted to create a good example of the new government policies "building a new socialist countryside" (*shehuizhuyi xinnongcun jianshe* 社会主义新农村建设) and "beautiful country buildings" (*meili xiangcun jianshe* 美丽乡村建设). Mr. Shen, a township official who was known to be good at organizing this kind of rural collective action, was sent to Mei Village in early 2013. He explained as follows:

> It was my job to make sure all the projects could move on successfully because it was government policy. However, normally, it was the village head who visited the related households and tried to convince them. I rarely talked to the villagers directly. When I was there [in the village], I organized the documents to apply for new projects. ... When we did something, we would invite some leaders to visit the village and we also tried to be reported on by the TV station or city newspaper, which would help us to get more projects. What I did was quite successful; the leadership was satisfied with my achievements. Mei Village could implement

3 A town mayor receives a stable salary from the financial department and is first nominated by the Communist party and then elected by the respective local People's Congresses. This is an indirect election based on the hierarchical electoral system in China.

> more projects. We built a public toilet and a public biogas plant, a road, drains, a dance stage, a basketball court, and a pavilion.
>
> interview in Mr. SHEN's office in Xintang Town, March 2015

When I conducted the survey in Mei Village in December 2014, I spoke to the village head, who pointed out that it was easy to obtain land from his relatives and some other households. However, two households were difficult to convince—even the mayor could not convince them to contribute their land for public use. The head of one household was a former cadre who could not obtain his pension after retiring. He claimed that the government had to give him his pension since he had worked for the township government from 1958 to 1994. However, the township government did not approve his application since he had lost his original documentation and the policy governing former workers had changed. Another household wanted to obtain compensation for contributing their land, but that is not part of the "One Project, One Discussion" system. Eventually, these two households were "forced" by the village head and the government as well as the rest of the villagers to contribute their land for a basketball court.

Although the village head and his relatives had also provided their land for public use, some cases of unfair distribution in later projects led to a crisis of trust. For example, the village head gave the benefits from a biogas project that was linked to the public toilet to most of his relatives, excluding others who had also contributed their land to this project. This unfair distribution resulted in public discussion and complaints. But the new construction projects were executed as scheduled and so were successful from a governmental point of view, even if they led to discontent and caused friction among the villagers.

3.2.2 Case Study 2: A 220-meter Cement Road Project in Mei Village in 2015

Construction of an additional 220 meters of cement road was planned for Mei Village in 2015. But the Qinbei District government had allotted funds for only 14 kilometers of cement road to the Xintang Township government. Why did the township government plan another road project in Mei Village, in addition to the 390-meter road of 2012?

I was told that this decision was made in order to complete "The Last Mile" (*zuihou yigongli* 最后一公里) of the projects in Mei Village (interview with Mr. Shen in his office in Xintang Town, March 2015). In the government narrative, "The Last Mile" stands for a planned project that could not be completely finished because of financial or management issues, usually related to corruption.

This terminology, of course, raises the issue of social and management problems in the future implementation of the project system.

The phrase "The Last Mile" is also regularly brought up by the government itself within the context of regular relocations of regional government leaders to other places, and in that context it refers to testing the consistency of government policies in the long term. Frequent changes in leadership or government mean that some important projects that are the responsibility of the higher-level government are prioritized, so that projects already in progress may be delayed and may not even be completed. As a result, such projects may languish with only "The Last Mile" to go before they are completed. In other words, this term reflects what Wegrich et al. (2017, 2) observed: "the delivery of infrastructure has become characterized by the 'iron law' … of cost overruns and delays." The term "The Last Mile" is also used in the context of eliminating the bad reputation of projects tainted by local corruption, but this is a goal that is not discussed openly. This problem is not unique to China. Time delays and cost overruns are commonly seen in Germany and other European countries, including cost overruns due to corruption (Kostka and Anzinger 2016; Fazekas and Tóth 2017). "The Last Mile" therefore is a Chinese response to a crucial and persistent issue within public infrastructure projects.

The Mei Village project, unfortunately, exemplifies the phenomenon of "The Last Mile." Figure 26 shows the government's blueprint for the road and the realities of road construction inside Mei Village.

The village head had already told me about the allocation of the 220-meter road in early 2015. There was a lack of trust between him and his fellow villagers. In March that year, the village head asked me whether I could do him a favor in Mei Village, which was to collect the contribution from each household for the new road. He said that the government wanted to complete the "last mile of road" inside the village. According to his calculations, he had to gather at least 3,200 yuan to fix the project. He had tried twice to inform the villagers about the road project and attempted to collect the money, but he had failed. I was curious why an elected village head could not collect money for a small road project, and thus I conducted several in-depth interviews and group discussions in Mei Village. At that time, a villager stated that he would not pay even a single penny for the new project; the same opinion was voiced by others.

I organized a discussion between three elders and other villagers, which showed that both the village head and the Village Finance Governing Group had lost the trust of many villagers during the former road project. People believed that if all of the profit earned from selling public sand had been spent on a 390-meter road, there must be something wrong. Nevertheless, both the

FIGURE 26 The Xintang township government's 2013 plan for a road in Mei Village (*a*) and the site of the proposed road in 2015 (*b*)
SOURCE: FIELD DATA, 2015

village head and the villagers wanted the next stretch of road to be built. The villagers pointed out that they needed a good road, but they also needed to know the financial details.

Thus, I organized a focus group to discuss the 220-meter road project and possible solutions to the issue. The people questioned the project system and its financial opacity. During many discussions, I realized that the best communication strategy would be for the township officials to clarify the project system and answer the villagers' questions face to face. First, I personally interviewed the township official, Mr. Cai, in his office in Xintang Town and explained the ongoing discussion. I then invited Mr. Cai and his colleague Mr. Feng to Mei Village to answer the villagers' questions about the road projects. I conducted these interviews to gain detailed information about the new road project, and I also wanted to help villagers to build the new road. The people agreed to contribute money to the new project, but only if I collected the money. This was one situation I encountered in which it was difficult for me to distinguish my position as a community member from my position as a researcher, as I discussed in the methodological section in chapter 2. I facilitated some of the communication and interaction between the township government and the villagers in order to build trust and enable further cooperation.

In 2015, the 220-meter cement road received funding and was constructed in four sections inside the village. Sixty meters of road connected the main road to the basketball court (by compulsory order of the Xintang Township Government), which was the first target of this project. The locations of the remaining 160 meters of road were decided by the village head, township officials, some of the villagers, and me. The second part was a further 20-meter stretch opposite the house of the village head, who had pointed out that whenever it rained, mud would clog the main road and muddy water would flow toward his house. The third part extended 120 meters from the village center to the east part where I was living. On my suggestion, the last 20 meters stretched from the basketball court to the house of the former cadre mentioned in the previous case study, to compensate for his lost pension and his involuntary contribution of land for the basketball court.

In both of these case studies, the decision-making process on the distribution of the roadway did not involve a mass meeting or elders' meeting. The signatures on the agreement concerning the 390-meter cement road project in Mei Village included some names that did not belong to villagers in Mei Village, and the village head said that he had added these names as instructed by local cadres. Indeed, the case of "The Last Mile" shows that policies surrounding the project system could be decided among the village head, myself, and local government officers, excluding the villagers. Even though public participation is the major purpose of the "One Project, One Discussion System" and village households contributed to the project fund, the "Last Mile" approach means that in reality, other systems are bypassed or not fully implemented. Friction between the village head and his fellow villagers was the product of these earlier projects. The villagers argued that "the money from sand selling was all used for the projects in a non-transparent way" (Mei villagers, March 2015). The village head responded: "The detailed cost and use were published in a notice posted in front of the shop. You can check anytime" (March 2015).

4 The Counterproductive Aspects of the Project System in Villages

It is worth mentioning that not all of the natural villages have their own cement roads; for instance, Naxia Village had none by 2017. Not every village has a running-water project either. Naxia and some other villages in Naqiu Administrative Village did not benefit at all from these projects during my research stay. And even though projects for water purification and an agricultural channel were implemented in Nachun Village and Naxia Village, the villagers were angry about the problems associated with these projects.

The principle of local people's participation is an outstanding characteristic of the "One Project, One Discussion" system. Nonetheless, failure to implement this principle in the project system is rather common. Apart from "One Project, One Discussion," most of the other projects are entirely unknown to the local villagers. In a group discussion in Naxia Village, villagers expressed their disappointment with the social entertainment facilities. They asked if I had seen an isolated building and a basketball court when I was driving toward the village. Then they explained that they were the fruits of a so-called benefit project to construct new social entertainment facilities in the administrative village, several kilometers from the residential area. Not one villager had agreed to the location proposed for these facilities, since it was too far away from all of the villages, and the facilities have indeed in general benefited none of the villagers. "We heard the project cost 2 million yuan in 2016," said one villager. "The Secretary of the CCP in Naqiu Administrate Village himself insisted on building there without giving a reason. People argued that given the location of the basketball court, the place for this building should be close to our village, but our voice does not matter. Until now, there is a ghost building and an empty basketball court" (group discussion in Naxia Village, March 29, 2017). I heard a similar statement in Nachun Village when I interviewed Zeng Pei and his mother on April 2, 2017.

Compared to the disgruntlement over the social entertainment facilities, water conservation projects and bridge projects are making people even angrier about government-funded projects. Interviews revealed that some of these projects had destroyed the original facilities without completing the new ones. For instance, the project for improving the roads to the farmland in Naxia Village was funded by the government. The company that obtained this contract would receive part of the funds. They first removed the previous road and irrigation channel and tried to build a cement one. However, the project stopped in the middle due to lack of funds or corruption. The interruption of the project badly disturbed the peasants' daily activities: the incomplete new roads and channel could not be used, while the old ones had been destroyed.

Another example is the Nadong Bridge project in Nadong Administrative Village. The original Nadong Bridge had been damaged by sand dredgers and trucks over the last three decades. A project to build a new bridge was started in 2015, and the previous bridge was destroyed. As of July 2020, however, the project was still incomplete—only three bridge pillars had been built (see Figure 27, a). The absence of a bridge disrupted the daily travel of more than 10,000 people who needed to cross the river, and both villagers and the local cadres complained a lot about this project. The "Last Mile" of the project was missing.

Irrigation channels and the agricultural water supply are other projects under debate. Zeng Pei and his two uncles told me that irrigation channels were only built where they could be seen from the major roads, so that leaders or outsiders would assume that the project was progressing. Such "vanity projects" (*mianzi gongcheng* 面子工程) are a big problem in government-sponsored construction work (interview in Nachun Village, March 30, 2017). The government has no interest in constructing agricultural channels in most of the paddy fields. A woman in Nachun Village said, "We do not have water that reaches my paddy field. The only working channel was built a long time ago and most of the households are struggling to get enough water to farm in Nachun Village" (interview, March 30, 2017).

Zeng Pei complained about a similar situation in Nachun Village. In 2017, he showed me that an alleged cement irrigation channel in a field was still the old mud channel (see Figure 27, b). He said,

> The cement agricultural road project was just a joke. After the conflict in 2015 [see chapter 7], some people put some paving gravel on the surface of the mud road. As you can see, the channel was a project costing several million yuan. However, it is just like a game played by kids.

FIGURE 27 The damaged bridge in Nadong Administrative Village (a) and the old mud irrigation channel in Nachun Village (b) in 2017
SOURCE: FIELD DATA, MARCH 2017

I asked him who put the gravel on the road.

> No one knows, because no one saw it. They did it in the middle of the night like a mouse. Why? Because they have been checking on the project. They are very afraid the officers from the higher level will visit the river, since there is a conflict beside our river, and discover the reality of the road project.
>
> interview with ZENG PEI, March 30, 2017, in Nachun Village

The case of the poor agricultural channel in Nachun Village is not unique. In Mei Village, I saw one peasant pumping water from the river in order to farm. "The channel has not worked for years," he said. A woman added, "The irrigation work is very bad. Furthermore, the water conservation was rented out to a private person for fishing. They save the water for fish farming rather than for rice growing" (Mei Village, April 8, 2017).

The "Renovation of Dangerous Rural Buildings" (*nongcun weifang gaizao* 农村危房改造) project aims to help poor households, as discussed above, but several circumstances have made it problematic. First of all, the village cadres benefited when people applied for funding. In a group discussion in Naxia Village on March 29, 2017, Mr. Hu said, "The secretary must obtain around 3,000–5,000 yuan from the total grant. Otherwise, he will not sign and seal the application document. He has the power to confirm whether a household meets the requirements." One elder said during the discussion that the secretary would not accept 3,000 yuan to sign the documents: "3,000 yuan was too little for him. He asks for more." In the case of a very poor family with an old traditional house and an ill family member, all of the villagers had agreed that this household should benefit from the "Renovation of Rural Dangerous Buildings" project. Nonetheless, the secretary refused to sign the application document, apparently because "the head of this household had contradicted him three decades ago" (group discussion in Naxia Village, March 29, 2017). Moreover, Mr. Nong has built cement agricultural channels for his production team while other teams' channels are still made of dirt or have been destroyed. Issues of corruption, personal affairs, and management capacity during the implementation of the projects have been reported on many times in the prefectural media, but officials continued to keep the majority of the benefits for themselves by embezzling from the project fund. Corruption has hindered the implementation of the project system and resulted in low-quality projects and distrust between villagers and the government in rural areas.

5 Conclusion

This chapter aimed to answer two questions: What causes the huge demand for river sand in China? And why has this natural resource been appropriated and turned into a commodity in the construction market? Building on the preceding analysis, I argue that the Chinese government's project system contributes to development and urbanization and is therefore increasing the demand for river sand. Rapid urbanization, increasing construction of public housing, and vast road and infrastructure construction projects are driven by governments at all levels. Thus, the government at various levels plays a crucial role in the commercialization of river sand.

The chapter also identifies counterproductive outcomes of the project-based development model. It furthermore illustrates the relationship between numerous development projects and interactions between the natural and social environment. By describing the ever-increasing investment in construction projects and the waste of sand, it shows the fundamental unsustainability of the era of project-based development. Instead of benefiting the local people, some projects damage existing infrastructure and erode people's trust in the projects' implementation. Incomplete projects reveal the problems and counterproductive nature of some aspects of the government project system, while exclusion of the public from decision-making goes against its initial purpose. Meanwhile, new roads and bridges built with sand from the villages are quickly damaged by trucks hauling more sand for other projects. Indeed, the weaknesses of the project system reveal the issues involved in project completion, transparency of funds, and the participation of local residents.

Thousands of projects are under construction. These projects spur growth in the building material market, causing a demand for thousands of tons of river sand. Rapid urbanization has consumed most of the river sand and has also led to a huge waste of river sand in the dismantled villages—especially when people build new houses that will speedily be demolished solely to gain more compensation. The unevenness of economic development within the province surfaces as an important topic for analysis when considering migrant workers and the translocal sand demand and market. Through the project system and river-sand marketization, the rural Zhuang communities are highly engaged in the process of rural modernization and labor–material mobility in the region.

In sum, the contribution of this chapter is to illustrate that China's developmental experience teaches a valuable lesson: the demand created by new development has exceeded the supply of raw resources. Furthermore, the model of rapid urbanization and project-based development has led to mismanagement of projects (e.g., the "Last Mile") and the waste of sand resources. Against this background, a detailed examination of the government's regulation of the sand industry is necessary to fully understand the sand crisis in China. The government's institutions and policies for controlling river-sand mining will be explored in the following chapter.

CHAPTER 5

State Ownership and Governmental Management of River Sand

The previous chapters discussed the drivers behind the changes in the demand for river sand, including the prominent role of state policies. However, the fundamental questions of who manages river sand extraction and by what mechanisms have remained unexplored, and I therefore address these questions in this chapter.

Laws and regulations have shaped and reshaped societies and the relationship between the human and non-human in many ways. Political ecologists have studied how laws provide legitimacy of access, control, and rights over resources and what role they play in resource conflicts. Laws express power structures and shape relationships and actions regarding resources (Goodale 2017). However, political ecology has focused much on land, a bit on water, and less on other resources, including sand. I have discussed sand's natural characteristics and its uses in the introduction. This chapter elaborates on how governance of sand is both similar to and different from governance of other resources.

This chapter is first and foremost based on debates in the field of political ecology concerning scale in relation to power and governance (Marston 2000; Neumann 2015; Smith 2000; Sheppard and McMaster 2004). A political ecology view attentive to scale provides a better understanding of the dynamics in which sand, as a resource, integrates with human desires, appropriation, control, benefits, and conflicts to shift resource regimes and social relations. The diverse levels reflect "differences of power, perception and of geographical distribution" (Kull and Rangan 2015, 487). Neumann (2015, 476) states that "attention to power asymmetries is critical for understanding networked relations within and between scales." This chapter therefore pays attention to power differences within and between the different levels of government involved in the management of sand.

By presenting the legal context first in the following section, this chapter focuses on the shortcomings of China's formal politics, both horizontal (institutions at the same level) and vertical (from the central government to administrative villages). In particular, political reforms and recently enacted regional laws will be discussed in detail. In order to control illegal sand mining and prevent contamination of water, China changed its sand-governance system from

 | DOI:10.1163/9789004505919_006

the license system (*caisha xuke zhidu* 采砂许可制度), which operated only on the county level and above, to the River Chief System (*hezhangzhi* 河长制), which operates down to the level of the administrative village. The reform of river-sand management has thus distributed responsibility downward, all the way to the level of administrative villages. This chapter explores the role of politics and the crucial part that government at various levels plays in resource management in rural China.

1 Laws Governing River-Sand Mining and the State Ownership of River Sand

Property rights are vital in the management of river sand and in its distribution, use patterns, and commercialization. If we study the Constitution of China (CPRC54 1954; CPRC04 2004) and the Property Law of the People's Republic of China passed in 2007 (NPCPRC07 2007) with regard to codified property rights, we see that there are three types of ownership: state, collective, and private ownership. But private "ownership" of land (individual-, household-, or company-owned) in China is really a private leasing of land from the state or from a collective; ownership remains vested in the state or the collective, such as a village (Ho 2001, 2015). Moreover, it is not clear how these forms of ownership operate in the context of economic development and legal implementation. The actual situation of multi-type common resources and land use makes the boundary between ownership rights and usufruct rights fuzzy. According to Ho (2015, 355),

> Despite theoretical foundations to the contrary, the country's capitalist growth over the approximate past three decades has occurred *without* formalized, legally protected property rights. In fact, it appears that the cadaster results from development rather than being an institutional precondition. (emphasis in original)

In practice, many resources are claimed by the state, while in rural areas, the majority of land and resources are claimed by communities collectively. As far as river sand is concerned, the constitution makes clear that all mineral resources belong to the state. The state council represents the country in governing the use of river sand.

There is no specific law on sand management in China. However, several laws on the management of other natural resources—such as legislation concerning water, the administration of land, and the management of rural land and of grasslands—are all relevant in this context.

Because of the ambiguous nature of river sand, it is subject to multiple and overlapping forms of governance, depending on the context. On the one hand, river sand is considered as part of river water, and its management has therefore been guided by the Water Law; for instance, river sand could be part of "water and soil conservation, prevention and control of sedimentation" (Article 14). On the other hand, when sand forms barren beaches, it is considered part of land, and its management can be governed by the Land Administration Law (Article 13). Meanwhile, river sand can also be seen as a mineral regulated by the Mineral Resources Law. Yet it differs from other minerals such as copper (copper miners and copper mines must be registered first according to law, and a mining license is required) since individuals and communities are allowed to mine river sand for personal or public purposes. Table 8 summarizes the stipulations in various national laws that together constitute a national legal framework for sand management.

The Water Law is the law that relates most directly to river-sand mining. Article 39 of the Water Law stipulates:

> Where sand quarrying in areas under river course control may adversely affect the stability of the river's condition or threaten the safety of the dykes, the administrative departments for water resources under the relevant people's governments at or above county level shall delimit no-mining areas or set no-mining periods, which they shall make known to the general public.

However, across the entirety of the Water Law, this is the only article relating to sand issues and simply deals with the state's institution of a licensing system for sand mining in rivers. Article 3 states that all water resources are owned by the state, yet it does not explicitly mention sand as a water resource. Nevertheless, the legal framework as summarized in Table 8 suggests that the ownership of sand lies with the state.

There are many other relevant laws impacting sand management, for instance the Environmental Protection Law of the People's Republic of China (EPLPRC89 1989, EPLPRC14 2014), the Regulations of the People's Republic of China on the Administration of the Channels of the People's Republic of China (SCPRC87 1987), and the Regulations of the People's Republic of China on River Administration (SCPRC88 1988). These laws attest that there are discrete authorities that have a voice in sand management and utilization.

TABLE 8 Laws regulating sand management

	Text	Year of issue	Adopting unit	Article(s)	Summary of relevant content
1	Water Law of the People's Republic of China	1988 (revised 2016)	The 24th Meeting of the Standing Committee of the Sixth National People's Congress	Article 24, 39	Article 39: "The State applies a licensing system for sand quarrying in river courses." (NPCPRC88, 1988)
2	Mineral Resources Law of the People's Republic of China	1986 (amended 1996)	The 15th Meeting of the Standing Committee of the Sixth National People's Congress	Most articles of the law	Article 3: "Mineral resources shall be owned by the state. The state ownership of mineral resources, either near the earth's surface or underground, shall not change with the ownership or right to the use of the land which the mineral resources are attached to. … Anyone who wishes to explore mineral resources shall register according to law. Anyone who wishes to exploit mineral resources shall apply for the right of mining. The state shall protect lawful rights of exploration and mining from violation and protect order in production and other work in the mining and exploration areas from interference and disruption." Article 43: "Anyone who, in violation of the provisions of this Law, purchases or sells mineral products which are to be purchased exclusively by the state shall have such products and his unlawful proceeds confiscated, and may be fined concurrently." (NPCPRC86 1986, NPCPRC 96 1996)

TABLE 8 Laws regulating sand management (*cont.*)

	Text	Year of issue	Adopting unit	Article(s)	Summary of relevant content
3	Rules for the Implementation of the Mineral Resources Law of the People's Republic of China	1994	The state council	Item 3 of the Mineral Resources Classification	Several types of sand are listed among the non-metallic minerals: natural quartz sand (glass sand, mold sand, building sand, cement ingredients with sand, cement standard sand, brick sand). (SCPRC94, 1994)
4	Flood Control Law of the People's Republic of China	1997 (revised 2007; amended 2016)	The 27th Meeting of the Standing Committee of the Eighth National People's Congress	Article 35	Article 35: "Within the scope of protection of flood control works, exploding, drilling wells, quarrying stones, collecting earth or other operations endangering the safety of flood control works are prohibited." (NPCPRC97, 1997)
5	Criminal Law of the People's Republic of China	1979 (revised 1997)	The Second Session of the Fifth National People's Congress	Article 343, Article 408	Article 343: "Whoever, in violation of the provisions of the Mineral Resources Law, mines without a mining license … if they refuse to stop mining after being ordered to do so, thus damaging the mineral resources, shall be sentenced to fixed-term imprisonment of not more than three years, criminal detention, or public surveillance and shall also, or only, be fined; if severe damage is caused to mineral resources, they shall be sentenced to fixed-term imprisonment of not less than three years but no more than seven years and shall also be fined." (NPCPRC79, 1979)

Item 3 of the 1994 Rules for the Implementation of the Mineral Resources Law stipulates that sand belongs to the category of non-metallic minerals. Therefore, Article 343 of the Criminal Law applies when it comes to river sand. It thus seems that sand is subject to a clear system of laws that regard it as a non-metallic mineral. In reality, though, China's environmental law and its implementation is highly problematic. According to Xin Chunying, the deputy director of the Legislative Affairs Commission of the Standing Committee of the NPC,

> Over the years, China has introduced about 30 laws on environmental protection, about 90 administrative laws and regulations, and a large number of environmental standards. It can be said that China's environmental problems are not due to the lack of rules, but because some of the rules have lagged behind practice. [These laws] are inoperable.
>
> quoted from PENG and MAO 2014

Van Rooij (2006a, 57) argues that although "China has made great strides in establishing environmental laws and enforcing environmental regulation, problems of non-compliance and weak and slow enforcement remain." Jin Wang (2010) states: "China's green laws are useless." Wang (2013, 367–368) adds: "Although China has constructed an expansive environmental law framework over the past 30 years, implementation of laws and regulations in practice has been notoriously weak. … Environmental targets (e.g., energy saving, emissions reduction), like China's environmental laws, had never been taken very seriously." Li and Shapiro (2020, 37) assert: "The legal system for enforcement remains weak." A number of researchers studying pollution explored the development of China's environmental conservation institutions and policies. This research demonstrates that although rather strict laws and regulations have been enacted over the past decades and even though these laws are enforced, the severe environmental crisis visibly persists (e.g., Jahiel 1998; Van Rooij 2006a; Wang 2013).

This situation is the same regarding the overlapping laws that are intended to prevent illegal river-sand mining and control legally permitted mining. These overlapping laws result in a fragmented approval system for legal sand mining and a lack of clarity on which formal institution is legally responsible.

Until today, existing laws regarding sand management have only rarely been effectuated, as the dominant factors in dealing with this resource are different. Meanwhile, there is no standard or consensus to guide the investigation of the criminal responsibility of illegal sand miners in China. The Regulations on Sand Mining Management of the Yangtze River were the first precise instructions directly related to sand-mining governance in China and were implemented on January 1, 2002. This is the origin of Decree No. 320 of the State Council of the People's Republic of China (R.-J. Zhu 2001). According to Decree No. 320, the Department of Water Administration of the local people's government at or above county level can manage river sand and issue sand-mining licenses. Thus, the township government and the administrative village as well as natural villages do not have authority over river sand. However, in addition to the Department of Water Administration, the Department of Land and Resources and the Department of Transport on the regional or local level also have authority over sand-mining issues. Various government documents and reports on combating the illegal mining and selling of river sand in China demonstrate that there are even more authorities that have local jurisdiction, including the Department of Environmental Protection, the Department of Agriculture, the Power Supply Bureau, and the Maritime Bureau. Indeed, not only do the laws overlap, but the specific authorities involved also vary according to place and the various governing systems, which are highly flexible and diverse.

In January 2015, the Supreme People's Court and the Ministry of Water Resources conducted a field survey on illegal mining and on criminal cases related to destructive mining in the Yangtze River and the Huai River Basin. On March 9, 2015, the Supreme People's Court proposed a draft interpretation of the findings of the joint survey, with the aim of soliciting opinions from the Supreme People's Procuratorate, the Ministry of Public Security, the Ministry of Land and Resources, and other departments. The draft brought to light serious instances of illegal sand mining that fall within the scope of criminal mining and destructive mining. But the process of drafting precise and effective laws on sand mining has not been completed yet, and there is still a long way to go toward implementing laws governing river sand in China.

2 Horizontal Politics in the Governance of River-Sand Mining

Horizontal governance refers to the interactions between different departments at the same level of government. The complex and overlapping nature of horizontal politics can cause serious management problems, as exemplified in the case of river sand.

Over the last two decades, many departments within the government on the county level and above have attended to the matter of illegal sand mining. Each relevant authority established its own policies on mining sand, and the common theme of these policies was the collection of a fee from approved sand mining. Thus, these departments utilized their power to benefit from sand governance. "Multiple management [and] indiscriminate digging and mining" (Wu 2008, 1–5) was the "method" used by some government authorities.

The Department of Water Administration under the State Council and the affiliated Yangtze River Water Conservancy Commission have strengthened the unified management, supervision, and inspection of sand mining along the Yangtze River, as well as organizing, coordinating, and directing the work of sand management. In March 2012, the Yangtze River Water Conservancy Commission established the country's first specialized, functional river-sand mining management department, the Yangtze River Sand Excavation Management Authority (*changjiang hedao caisha guanliju* 长江河道采砂管理局). However, the newly established authority has not succeeded in controlling illegal sand mining, while many other powerful authorities desire to get involved in sand governance.

The Yangtze River Sand Excavation Management Authority organized its cadres and pointed out this situation in a book it published in 2008:

> Despite the implementation of Decree No. 320, which clarifies the management system for river-sand mining and explains the division of responsibilities between the water administrative department, traffic department, and police, some departments do not fulfil their duties in actual practice. They do not want to give up pursuing their interests. They use their own industrial authority, trying to issue some permits for sand-mining operators. In so doing, they pursue economic interests regardless of the overall management of sand mining. Thus, they are increasing the costs of sand-mining operations and sand-mining management, as well as the difficulty of on-site supervision of sand mining.
>
> WU 2008, 61

Government departments operating at different scales have frequently established joint campaigns or joint law enforcement activities in order to form a stronger team and take united action against illegal sand mining. Liu et al. (2010) argue that ecological issues in general straddle administrative boundaries (e.g., central, provincial, and prefectural), and formal governance can therefore benefit from such cooperation. However, Li and Shapiro (2020, 55) analyze the effect of the joint political campaigns as follows:

> Through these campaigns, the Chinese state recoups both consolidated control over its local counterparts and political legitimacy in the eyes of the people. As we have shown, environmental campaigns offer short-term rather than lasting gains and all too often snare law-abiding citizens

> in their nets. Campaigns create the appearance of the state's strong environmental will, but they gloss over problems such as non-compliance, inequality, negative externalities, and arbitrary enforcement. In the end, campaigns produce an ad hoc patchwork, dashing hopes for extended sustainability.

When law-enforcement bureaus joined forces to destroy the illegal sand-mining sites and machines around the Maoling River in the jurisdiction of Qinzhou, media outlets reported on their actions and revealed how illegal sand mining has been policed:

> From November 9 to 10, 2017, for two consecutive days, Qinnan District organized some 900 officials from 23 departments including the City Supervision Bureau, the Land and Resources Bureau, the Water Resources Bureau, the Marine Bureau, and the Environmental Protection Bureau in Qinnan Government District to comprehensively manage illegal sand mining and illegal operations in sand fields on the banks of the Maoling River. [The joint campaign] vigorously combats illegal sand extraction and illegal operations in the sand fields to maintain the ecological environment.
>
> In recent years, with the rapid economic and social development of the city, the market demand for river sand and sea sand is increasing over time, providing huge profit margins for illegal miners. In addition, the legal awareness of some people is weak, and they consider extracting river and sea sand normal. The illegal operation of sand-mining and-selling sites is relatively common and has caused great damage to the ecological environment. In order to protect the ecological environment and to ban illegal and random extraction of river sand and sea sand and the operation of sand-selling sites, Qinnan District has repeatedly organized law enforcement officers to crack down on illegal sand mining in the areas of Jianshan Town, Nali Town, and Kangxiling Town.
>
> QINZHOU360 2017

Similar efforts to control sand mining have been reported in the Zhaoqing City section of the West River in Guangdong Province:

> In 2016, 700 officials instigated a campaign against illegal sand miners and illegal selling sites around the river. According to incomplete statistics, over 90 illegal river-sand storage yards were found along the West River in Zhaoqing City in May 2016. Forty-five people who had entered

> the illegal sand-mining business were arrested by 150 police officers on March 27, 2016.
>
> XIAO and TANG 2016

As Jia Shihao, a member of the Prefectural Security as well as the official in charge of the campaign, pointed out,

> An ordinary illegal sand-mining ship can harvest 1,000 cubic meters of river sand in 20 minutes. In one night of normal operation, it pumps between 1 and 1.5 million cubic meters of river sand. According to the current market price, which ranges from 60 yuan to 120 yuan per cubic meter of sand, it will gross 600,000 yuan to 800,000 yuan each night, and even minus costs, it still can earn three to four hundred thousand yuan.
>
> XIAO and TANG 2016

The number of departments and officers involved in these policing efforts not only shows how rampant illegal river-sand mining is, but it also illustrates how overlapping responsibilities and fragmented sand management pose problems for government agencies on the horizontal level. The overlapping authorities have a problem managing river sand in an effective way. Consequently, local governments have been blamed for unfaithfully implementing the central government's wise environmental policies in China (B. Van Rooij 2006a; Mol and Carter 2006; Kostka and Hobbs 2013). However, changes in the central policy regarding which authorities are actually responsible for overseeing sand mining are an important root of sand governance problems. From 1970 to 1990, the Ministry of Land and Resources was considered the most important authority when it came to sand management. But the responsible departments at the state level have changed over the years, from management by the Department of Land and Minerals, to joint management with the Department of Water Resources, and then to management by the Department of Water Resources alone. These changes in jurisdiction illustrate how the perception of river sand has changed over time. Originally it was seen as an unlimited mineral resource that could be appropriated at the discretion of local authorities as well as individuals; most recently, river sand is considered an inherent part of the river environment that plays an important role in the ecology of the river system and in the sustainable development of rivers (Wu 2008). This changing perception regarding the classification of sand as a land resource or a water resource explains how ministries within the Chinese government change their ideas on the management of river sand.

In actuality, the Department of Water Resources has difficulties governing sand mining without the help of other authorities. On the one hand, the benefits and power associated with the management of a valuable resource have led to department protectionism. As Van Rooij (2006b, 4) explains:

> Local protectionism [means] that local governments let their own local interests prevail over national concerns. Local governments were able to exercise such protectionism through their control over the budgets and personnel management of their local bureaucracies—including courts, prosecutors, police, and administrative departments such as land bureaus, industrial bureaus and environmental protection bureaus.

Wu (2008) pointed out that department protectionism, fragmented approval systems, and blind spots in laws and regulations are the major issues in river-sand governance. On the other hand, the boom in river-sand mining has put tremendous pressure on the river ecosystem and on flood prevention. The lack of law enforcement inspectors and law enforcement tools faced by the government at various levels has been reported in the media.

A further issue is simply that there are a multitude of rivers in China, and sand mining is taking place in so many of them. Serang Yeshe, an official with the government of Baiyu County on the Jinshajiang River, a major tributary of the Yangtze, told *Xinhua Daily* on August 8, 2017: "There are 230 rivers in Baiyu, and there were many illegal mines along the rivers. … We only had five law enforcement inspectors. We increased the number to 15, and trained 30 people to assist with inspections." He further pointed out, "If we allow the illegal mines to continue, there may be a great risk of man-made disasters in the flood season" (Meng 2017). Thus, in Baiyu County and many other locations, a lack of law enforcement personnel led to a failure to meet the goal of effectively managing sand-mining activities, such as monitoring the time and duration of mining, supervising the permitted extraction of a certain amount of river sand per day, and evaluating the technical skills of workers and the technological requirements of the dredgers. As a result, the government failed to ensure that sand miners improved environmental protection facilities to meet sewage standards or to meet the legal requirements for mining installations. Due to these problems, some officials proposed to introduce new technology to combat illegal activities, such as drones and big data (Cheng 2018).

An additional problem is that the "devolution of fiscal responsibilities made horizontal government relationships stronger than vertical ones, such that a county government has greater control over a county environmental protection bureau than does the provincial environmental protection office" (Yeh

2015, 625; for similar arguments, see Jahiel 1998; Lieberthal 1997). The issue of vertical politics will be elaborated in the following sections.

3 Vertical Politics in the Governance of River-Sand Mining

Natural resource management in China is crucially dependent on governmental structure. The government has used "hierarchical management" (a top-down government structure), "administrative jurisdiction management" (based on spatially limited administrative boundaries), and "classification management" (differentiated according to various elements or features of the landscape, such as water management, land management, and forest management), as well as a series of other institutionalized administrative systems and actual operational mechanisms to achieve environmental and resource management (Han 2012; Wu and Chang 2020, 217; Ma et al. 2018, 53; Wu 2008).

Since Deng Xiaoping took power in the 1970s, economic decentralization (fiscal decentralization) and political decentralization (administrative decentralization) have played a crucial role in China's political reform to achieve economic development (Landry 2008; Montinola, Qian, and Weingast 1995). Landry (2008) argues that authoritarianism and decentralization can go hand in hand during political reform to decentralize decision making and fiscal revenue, on the grounds of economic efficiency. But the key factor to maintaining the authoritarian system is through the personnel management system of the CCP. Zhili Bai and Juan Liu (2020, 65) argue:

> China's governance model is a dual structure model that can be observed in its organization, decentralization, value and efficiency systems. At the same time, this reform, marked by the reform of Party and state institutions, establishes a party-centered state governance system and structure; that is, the adjustment of the relationship between politics and administration, promotes the construction of the state governance system of party centralism.

Nonetheless, river sand has become a cause for concern, triggering the profound transformation of related policies and institutions. Scale is crucial and must feature as a major dimension in any attempt to analyze China's complex governance structure.

According to Decree No. 320, the Department of Water Administration at or above the county level can manage river sand and issue sand-mining licenses. While the decree specifically applies to the Yangtze River, other governments

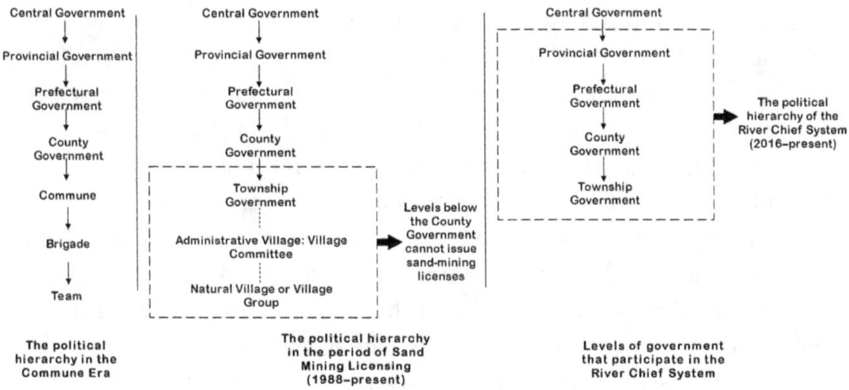

FIGURE 28 The three types of resource governance in China
SOURCE: DRAWN BY THE AUTHOR, BASED ON LUO 2014; DECREE NO. 320, 2001; AND GOSCPRC 2016, RESPECTIVELY

on the provincial level and below, based on their specific situations, must adopt and establish their own sand-mining regulations in their respective political territories.[1] The licensing system for sand mining was put into effect in this formal government structure, which is a form of vertical management. Thus, township and village-level administrations are excluded from the formal sand-governance structure. This is very different from the government's usual top-down structure, which extends down to the administrative village level.

Government structures have considerably changed in the last decades, as illustrated in Figure 28. The left part of the figure shows that during the commune era, from 1958 to 1984, sand mining was governed within the general political hierarchy established under the leadership of Mao Zedong. The central part of the figure displays the political hierarchy of the sand-mining licensing system that was introduced in the Water Law of 1988. The right part of the figure shows the political hierarchy of the River Chief System. Due to the ineffectiveness of the license system in governing water resources and controlling illegal river-sand mining, the central government continued to

1 According to Article 39 of the Water Law of the People's Republic of China, the state must implement a license system for river-sand mining. According to Articles 4 and 5 of the People's Republic of China River Management Regulations, the water conservancy administrative departments are the competent authorities regarding rivers in the administrative region. Article 3 of the River Sand Mining Management Regulations of the GZAR stipulates that the water administrative department at or above the county level is responsible for unified planning, administrative licensing and supervision, and management of river-sand mining.

reform river-health and river-sand governance. The central government's comprehensive reform leadership group reviewed and approved the "Opinions on the Full Implementation of the River Chief System" (*guanyu quanmian tuixing hezhangzhi de yijian* 关于全面推行河长制的意见) (Liao, Zhong, and Fu 2018), and this document was issued by the General Office of the CCP Central Committee and the Office of the State Council at the end of 2016.[2]

The River Chief System is a political reform based on the principle that the top leaders at each level of government from the province to the township are the "chiefs of the river" who are in charge of and responsible for the river (Hui Liu et al. 2019; Huang and Xu 2019). The government planned to build an administrative system as well as an assessment system to evaluate the effectiveness of river management. The "Opinions on the Full Implementation of the River Chief System" states:

> The party and government leaders served as the chiefs of the river, implemented the main local responsibility according to the law, coordinated and integrated the forces of all parties, and effectively promoted the protection of water resources, the management of waterfront shorelines, the prevention and control of water pollution, and the management of the aquatic environment.
>
> GOSCPRC 2016

One of the major goals of this system is to "resolutely crack down on illegal activities involving rivers and lakes and to resolutely clean up illegal sand mining activities" (GOCCPCC and GOSC 2017).

By 2018, the Minister of Water Resources reported that over a million river chiefs had been appointed, from the provincial level down to the administrative village level (MWR 2018). This is quite similar to a top-down cadre evaluation system, applied to water quality management (Golding 2011). It is also similar to a measure that assigned pollution reduction and energy efficiency targets to local cadres during the 11th FYP (2006–2010). In fact, as Wang (2013, 368) notes, " 'Environmental cadre evaluation' has since become the central focus of China's domestic and international environmental strategy. Central

2 The "Opinions" are effectively a proposal of a national policy. All governments at and below the provincial level must obey and establish a discrete River Chief System based on this document. As Myšička (2009, 62) notes, "Central government nowadays determines only general guidelines of state policy; however, it controls personnel policy of all administrative structures."

authorities have applied the tool to a range of new priorities concerning climate change, energy efficiency, and pollution."

This reform shows that China's government realizes there is a crisis in river-sand mining, water pollution, and degradation of the ecosystem of rivers and lakes by highlighting the responsibility of local leaders, who are also the officials in charge of river management (see also Liu et al. 2019b; Huang and Xu 2019). A top official serving as a river chief who failed to manage a river effectively and sustainably would probably jeopardize his or her political career.

The reform seems to be a response to the disagreements between the central government and local governments over who is responsible for river management. But the River Chief System is limited in scale, as Chien and Hong (2017, 65–66) underline:

> Even though the river leader policy is effective in the single dimension of pollution improvement over the short term, we also notice that the CCP has not introduced the policy to all rivers within China. While the central government in Beijing requires provincial governments to establish river leader systems for regional rivers and lakes across prefecture-level and county-level administrations within their provincial jurisdictions, there are no river leader systems for major rivers that run across multiple provinces such as the Yangtze River or the Yellow River.

Likewise, rivers within one county and small local basins also fall outside of the provincial River Chief System (Chien and Hong, 2017). Figure 28 reveals that the normal government power structure was reestablished regarding river-sand mining and water management according to the River Chief System in 2016, at the level of the province / autonomous region, prefecture/municipality, county, and township level governments (a four-level system), but many provinces have included village-level governments in the new system (a five-level system), such as Guangdong Province and GZAR. Although the sand-mining license system is not administered below the level of the township, the River Chief System shows that the government has strengthened its executive power for all levels right down to the village.

To implement the political reform, Qinzhou City established its River Chief System (Figure 29) on July 27, 2017 (Qinzhou Water Conservancy Bureau Office 2017). The city appointed government leaders at the prefectural level and below. There are two river chiefs at the prefectural level and 2,169 river chiefs at the administrative village level. Maoling River is not governed by the prefectural government but by the regional government, since it passes through two cities (Qinzhou City and Fangchenggang City).

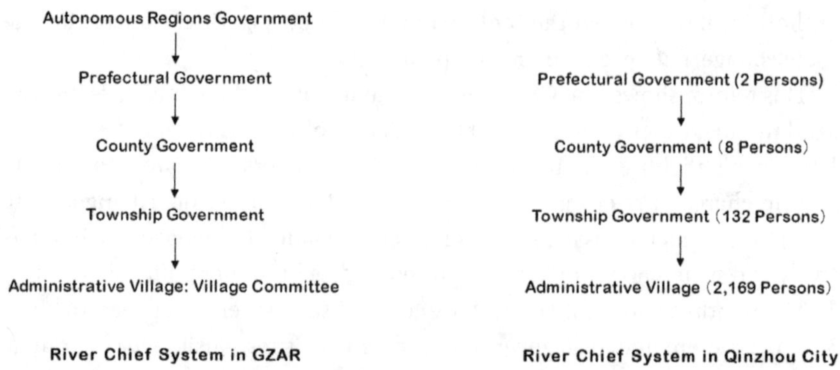

FIGURE 29 The River Chief System in the GZAR and in Qinzhou City
SOURCE: DRAWN BY THE AUTHOR, BASED ON GOGZARG 2017, RCOQ 2017, AND XIA 2018

The River Chief System relies on investigations and surveys, symposiums, and visits to the river as the major methods for monitoring the river situation. The frequency of implementation is diverse; the river chiefs of the prefecture execute their duty at least once each year, while the village-level river chief must inspect the river at least every week.

In addition to the River Chief System, the central government also sends the work team of the Central Environmental Inspector to each province and prefecture. Scholars report that interventions by the central inspector have contradictory effects. On the one hand, local government officials are liable to take extreme measures due to top-down political pressure. As argued by Li and Shapiro (2020, 55), "face[d] with tremendous political pressure from top-down inspections, local officials sometimes take it upon themselves to over-enforce environmental regulations by closing all factories in an entire sector." On the other hand, Van Rooij (2006a, 59) argues that "lax enforcement is rooted in conflicts of interest between national regulations and local stakeholders. As a result, the law lacks local legitimacy and local actors resist enforcement." The politically motivated are not working well with the reality of river sand management in the Zhuang communities. Over-enforcement is extremely difficult to maintain due to the high cost, while lax enforcement becomes increasingly common. I will discuss the consequences in a later section of this chapter.

The tension between the central and local governments in China has been well researched (Kostka and Nahm 2017; Edin 2003a, 2003b). Many of these authors argue that either the central government or the local government should be blamed for the failure of environmental governance. Qi and Zhang (2014) argue that local environmental enforcement is constrained by central–local relations in China. Ran (2013), on the other hand, argues that instead of local governments, the central government should be blamed for environmental degradation due to its failure to provide sufficient incentives to motivate local officials to enforce policies to achieve the overall conservation targets. The "perversely structured incentive system" is in Ran's view the reason that problems are continuing in local environmental politics. Eaton and Kostka (2018) in their turn call for attention to local–local relations.

The political reality in China is that local governments have been blamed for failing to implement the well-conceived environmental policies of the central government. Mr. Wu, the director of the Yangtze River Sand Excavation Authority, wrote:

> Other main issues include the failure to curb local protectionism, even though the Yangtze River's Sand Management Regulations implemented a local chief executive responsibility system and asked Hubei Province, Jiangxi Province, Anhui Province, and Jiangsu Province to sign a sand-mining management responsibility agreement (*zerenzhuang* 责任状) with subordinate governments. However, due to the high profits from river-sand mining, some provinces, cities, counties, and villages take sand mining as a major economic industry and even the pillar industry for development. Local protectionism includes the high-volume auctioning of sand-mining management rights, a multifaceted collection of taxes and fees, relaxation of sand mining supervision, weak enforcement of sand institutions, and a one-sided pursuit of maximum benefits. The central government has not established a system of accountability, which makes it difficult to really establish the responsibility of the local administrative heads. Therefore, the system for ensuring the responsibility of local people who act as executive heads is ineffective.
>
> WU 2008, 60–61

Most rivers are located in rural areas where the local government has profited from the excavation of river sand and protectionism is rife. Nevertheless, the research findings show that analyses based on top-down or central–local frameworks alone are too narrow to fully apprehend the problem of sand overexploitation, since the vertical axis is dynamic and local communities and villagers have been ignored by this formal structure. This is so in the case of the Maoling River: ever since the River Chief System was implemented in 2018 and the prefectural and county governments jointly extended their governance and monitoring down to the administrative village level, the river sand of Huangwutun Town has been sold exclusively to one company and the villagers have no voice in the matter.

Notably, there were still illegal river-sand-mining and sand-selling sites near the Maoling River in 2019, as I was told by an official of the Qinnan District government (interview on WeChat, August 10, 2019).[3] I asked him why there are still illegal miners after the implementation of the River Chief System. He replied: "Does the chief of the river have power? No." River governance is just a small part of the work of the officials who have been appointed as river chiefs. This means that the river chiefs do not have real power to control illegal sand mining. The following section will further outline the position of the provincial government vis-à-vis the central government when it comes to sand management.

4 Sand-Management Policies at the Provincial Level: Power Decentralization and Autonomy

The principles of power decentralization and of territorial management by local governments allow each provincial government to enact its own laws and policies and enable local officials to choose which national policies will be implemented and to what extent (O'Brien and Li 2006; Economy 2004). For instance, the Guangdong provincial government first adhered to national regulations on sand mining in the Yangtze River, but in 2005 it enacted its own provincial law. This supports some scholars' observations that in China, a local government has great authority over environmental management in its political territory (Landry 2008; Mei and Pearson 2014). Zhili Bai and Juan Liu (2020, 6–7) point out:

> China's economic development model is characteristic of the local government-led model of economic development [and] local party committees and governments in China have played an irreplaceable role in the rapid development of China's economy. This local government-led economic development model in China is attributed to the emergence of a more significant vertical decentralization pattern in the state governance structure.

However, the capacity of local governments to manage activities in their territories varies from region to region in China. For example, only recently did the

3 According to the Huangwutun Town government report of March 2019, 75 illegal sand-mining sites were demolished due to a report of the Central Environmental Inspector. Two chiefs at the township level shared the overall responsibility for sand mining in the township. There were also 12 other river chiefs at the township level and 23 chiefs at the village level.

government of the GZAR enact its regional law for managing river sand. Due to the rampant illegal river-sand mining taking place in many cities and towns of the GZAR, the regional government realized there was a need for a regional law on sand-mining management. On November 30, 2016, eleven years after Guangdong Province enacted its sand-mining legislation, the 12th People's Congress Standing Committee of the GZAR adopted its first regional law on river-sand management. The regulations were implemented on January 1, 2017. The Autonomous Regional People's Congress explained in press conferences on December 1, 2016, that the aim of the law was to improve the effectiveness of sand management and to combat illegal sand-mining activities.

At first sight, the recently enacted regional laws and the latest central political reform, the River Chief System, have similar purposes. However, a closer look reveals differences between the two. The party-state policy highlights the leaders' political responsibility for river health in order to achieve the desired goal of President Xi: "to implement the strictest possible systems for environmental protection, and develop eco-friendly growth models and ways of life," as declared during the 19th CCP National Congress in 2017. The provincial government considers the weakening of bureaucratic river-sand management as the most pressing problem in relation to river health and is trying to enact new laws with increased fines in order to receive more income.

Importantly, one of the most outstanding characteristics of the provincial regulations is the recognition of the sand-use rights of rural residents. The River Sand Mining Management Regulations of the GZAR (2016, Article 35) state that rural residents who only harvest a small amount of sand outside the prohibited areas are exempt from obtaining a mineral sand license and from paying sand-mining fees (SCPCGZAR 2016).[4] Yet, this law causes many disputes and has provoked many struggles relating to sand governance in the villages. An example of disputes between residents of local villages and the sand-mining company licensed to operate in those villages will be given in the following section.

5 River-Sand Auctions at the Prefectural Level: Sand-Mining Licensing in Practice

As in other cities, the Qinzhou prefectural government is facing rapid increases in demand for sand for construction and must respond to increases

4 A similar stipulation can also be found in some other provincial sand-mining regulations, such as in the first regional regulations on sand-mining to be enacted in China, in Guangdong Province, which allow households to extract sand for their own use but not to sell it.

in sand-mining activities in various rivers and coastal areas in its territory. The government's sand excavation policy has incorporated auctions into its sand-mining license system. In the following, I first describe river-sand auctions in Huangwutun Town to illustrate how the sand-mining license system functions in Qinzhou City.[5] Such auctions have been conducted by governments at or above the county level to sell the sand-mining rights for certain parts of a river in Qinzhou City. Issuing mining licenses to certain enterprises is a way to commodify river sand and increase the government's revenue. The weaknesses of the sand-mining licensing system on the prefectural level are related to overlapping interests, mismanagement, and corruption, which I discuss in more detail in section 5.2.

5.1 *The Deceptive Success of Auctions of Sand-Mining Licenses*

On February 3, 2008, the government of Qinzhou City issued an official announcement addressed to the governments of Qinnan District, Qinbei District, and other regional units under the scope of the Qinzhou government. Titled "The Sand Mining Rights Transfer Program of River Sections of Qin River, Maoling River, and Dafeng River," the document announced sand auctions in the city and focused on how to disseminate information about these auctions to the public. The document had been printed and issued to the relevant authorities, which were then consulted in detail about the organization of the auctions and mining management; later, it was published in the *Qinzhou Daily* (*Qinzhou ribao* 钦州日报). The document specified that two levels of government (prefectural and county governments) would be involved in the auctions and would share the income from the auctions. The point of the announcement was to clarify the management duty of each relevant agency in the two levels of government to make sure that mining would proceed smoothly.

The prefectural government sold sand-mining rights in five different sections of three different rivers between 2008 and 2010, including a section of the Maoling River passing through Huangwutun Town, which yielded a huge profit. Under this system, the government shifts its managerial tasks and law enforcement responsibilities to the mining companies, which are responsible for hiring people to prevent unlicensed sand businesses, dredger owners, or local villagers from mining their sand. (QZGOV 2008) Thus, the government spends less on management than it would if it were directly combating illegal sand mining, while at the same time collecting high profits from the license

5 In Xintang Town, the prefectural government and the district government have never been involved in the management of river sand because the river there is rather small. Hence, there was no auction in that town.

auctions. The cost of law enforcement in the city is always paid by the sand companies; some companies hire local criminal gangs to reduce this cost, since the government's fee for protecting the rights of sand companies is higher than that of a local gang, but the outcome is the same.

As noted earlier, the local officials and sand-business people call the auction winners *lau lao*, which means people who speak Mandarin but not the local dialect or Cantonese. Government officials in Huangwutun still called the license holders *lau lao* when I interviewed them in 2017. The *lau lao* became the "big bosses" of the whole river section in Huangwutun Town. They were originally from Zhejiang Province and Hubei Province, and they brought big new dredgers to the Maoling River to mine sand. Mining efficiency significantly improved and the price of sand increased, as discussed in chapter 4.

Unfortunately, rapidly escalating conflicts emerged between the auction winners and unlicensed, often small, companies owned by locals. The illegal operators did not have enough money to attend the auctions, as they were only open to companies with a minimum registered capital of 500,000 yuan or more. However, they had invested money to buy dredgers and other machinery and had earned a considerable profit from sand mining before the auction. The government auctions harmed their profitable business. When I interviewed an owner of an unlicensed sand-selling site, he told me:

> There were fights every single day [after the auction]. The government sent armed policemen to help the auction winner and to try to stop us from mining. The government hired excavators to destroy our sand-selling site and confiscated our belongings in 2008. For self-protection, we organized our relatives and fellows to go up against them [i.e., the government and the company that won the auction].
>
> interview at the owner's sand-selling site in Huangwutun Town, April 10, 2017

But soon, both sides realized that the conflicts did not help either party to make profits, so they decided to collaborate. Local small companies could buy permits from the successful bidders (e.g., Guang Xi Longtai Power Investment Company, at that time) and pay for the right to extract sand. This would cost 7 yuan for a cubic meter, one informant told me.

Apart from sand-mining companies, residents of villages located close to the river also wanted to extract sand for private use. However, the *lau lao* claimed that the sand belonged to their company and that the villagers had no right to extract sand anymore, except when they paid for it. Nonetheless, according to the River Sand Mining Management Regulations of the GZAR, the village's households have the right to harvest sand for personal use. The villagers therefore united against licensed sand mining in local areas. A former teacher at the middle school of Nameng Town told me in 2015 that in the previous year, the riverfront residents had stood up against a businessman who wanted to start mining sand in Nameng Town, even though he possessed a sand-mining license from the government. The businessman had already owned the license for a year and could not start extracting sand. The tension between the company and the village was high; disputes were increasing between these diverse actors, and the government did not know how to deal with the conflicts.

Indeed, legitimacy became a big problem. From the perspective of the prefectural government, the one who pays the fee and thus obtains the sand-mining license is the legal operator. By contrast, the village challenges these "legal operators" because sand-mining activities have damaged their land. Villagers highlight that "the state protects the farmland" and use the land-protection policy of the central state to argue against the prefectural government's licenses. Furthermore, villagers claim that the sand belongs to them as a village-owned property, while others argue that appropriating sand for personal use is legal due to the Mineral Resources Law and several regional regulations. However, the licensed companies insist that individual sand miners (whether they are households, villages, or small sand dredgers) are illegal, because the companies are the only legal subjects when it comes to sand mining according to their contracts with the government agency. The government, which has to support the legal mining company based on the sand-mining system, is thus in collusion with the sand-mining operators and has no way to deal with villagers' appeals.

Finally, some operators without sand licenses or permits from the companies also dredged in the very stretch of river that had been sold to the contracted company. And so legal and illegal companies were harvesting sand in the same river section. Thus, over the last two decades the local government and legal sand miners have not been able to eliminate illegal ones.

Concerning the sand auctions, one officer of Qinbei District, Mr. Lu, told me that it is mainly the villagers who feel the environmental consequences of legal sand mining:

> All in all, the major disputes revolve around the issue that the overextraction of sand has harmed rural villagers. Some people said that no matter how much the sand companies paid the government, sand mining was still forbidden in their river. Even if government officials went to mediate the conflict, the local people stuck to their arguments and protected their rights.
>
> interview on WeChat, June 8, 2016

These sand-governance disputes reveal inherent contradictions between sand policies, institutions, and laws. They show that these laws, institutions, and policies have not contributed to solving disputes involved in river-sand mining, but rather have caused new conflicts.

The Qinzhou Prefectural Government issued a notice on strengthening the management of river-sand mining in 2008, followed by a work program in August 2009. The work program aimed to counteract illegal mining and to regulate the mining activities of license holders in those sections of the rivers for which sand-mining rights had been auctioned. In the notice, the government explicitly welcomed the people's participation in sand-mining governance. It established a hotline and a reward system for reporting on illegal mining. Moreover, the work program was presented to all governmental departments involved in sand-mining management, to clarify their respective duties and responsibilities.

For instance, the Qinzhou Power Supply Bureau was charged with monitoring the electric power usage of river-sand quarries and cutting off electricity to illegal sand quarries. The police were to take action against the "sand-head fee" or "snake-head fee," which was collected by local villagers and other "gangsters" (introduced in Chapter 3). The district governments and the Prefectural Land and Resource Department were supposed to create detailed documentation of the sand-selling sites, for instance by videotaping the riverbanks and monitoring erosion, and they were also responsible for routine supervision of the area as well as its inspection and management. Finally, the Traffic Department was supposed to prosecute illegal sand transportation on the main roads.

Despite the well-drafted work plan, the Qinzhou government failed to monitor the companies that obtained sand-mining licenses due to the high cost of monitoring systems as well as its own conflict of interests in sand mining. The companies had paid a high fee to the government to purchase the licenses by auction. In order to recoup the costs and make a profit as soon as possible, they disobeyed the contracts, which clearly regulated the time periods, number of dredgers, and areas along the river where sand extraction was allowed. In effect, the government tolerated the companies' disobedience because of the income it received from selling licenses. The licensed companies were at the same time complaining that the government had failed to stop other illegal miners from harvesting sand in the very river sections that were sold to certain companies. Many private, unlicensed dredger owners in Qinzhou City continued to extract sand, indicating that the government was taking insufficient measures to punish these "thieves."

In short, instead of bringing peace and lawful sand governance to local areas, the license system brought complaints, petitions, and disputes.

Consequently, sand-mining companies took little interest in subsequent auctions. In December 2014, the government of Qinnan District planned to auction two sections of Maoling River and Dafeng River in the district (Project Number: KLQZG20141064). But the Qinzhou Finance Bureau later reported that the auction would not take place because it had not attracted three bidders, the minimum number required by law.

Nonetheless, the strong market demand for river sand encourages the government to involve itself with this profitable resource. The government at many levels and in various places remains active in the marketization of river sand. In October 2018, the Maoling River and Qin River in Qinbei District had been auctioned to the government-owned company Qinzhou Huangma Resources Development Investment. This company obtained the sand-mining management rights to several sections of the rivers and attempted to sublet its mining rights to other companies by posting a sand-mining partner recruitment announcement on social media as well as conventional outlets. The company was trying to sell its mining rights to thirteen sections of the two rivers, with a total length of 35,669 km and a total area of 1,670,000 m^3, for at least 835 million yuan (interview with an official of Qinbei District Government, January 2019; see also Huangma Assets 2019), indicating the massive size of the river sections up for auction.

Thus, the sand-mining process in Qinzhou City and its districts is always accompanied by disputes and conflicts. Moreover, the different levels and departments of government involved have not been able to stop unlicensed sand mining in the area.

5.2 *The Effects of Corruption and Mismanagement on the Sand-Mining License System*

The role of local authority is crucial in sand governance, as illustrated above. "Does policy interpretation balance local expectations with official views? Are officials subject to pressure from local elites, corporations or peasant organizations? What role does corruption play in law and policy implementation?" These are some questions that Bryant (1992, 19) asked when he explored the role of state policy and its implementation in developing countries from a political ecology perspective. These questions apply to river-sand mining policy in China; in particular, corruption of local officials has been one of the most serious issues in the sand-mining license system. Discussing the transition of China's governance model and system, Zhili Bai and Juan Liu (2020, 9) state:

> The most significant problems should be the three profound policy themes of environmental issues, corruption and income disparity that

> have been coexisting with or produced by China's long-term high-speed economic development since the reform and opening up. Anti-corruption reform could be said to be the deepest and most productive area of current reform. It is closely linked to the occurrence of environmental problems and income gaps, which is also related to the government-led or local government-led economic development model.

Due to a lack of law-based sand mining governance, illegal sand mining is an administrative rather than a criminal issue in many areas in China, and the most severe penalties are fines and a seizure of illegal sand-mining equipment. A large number of departments involved in sand management collect fees and fines. "They come for money, [they have] no other purposes" is how miners talked about the goals of governmental sand management in Huangwutun Town. Indeed, departmental protectionism has resulted in a situation in which there can be up to 19 different kinds of fees for a single sand-mining site in some places (Wu 2008, 6–7).

The maximum fine for illegal sand-mining defined in the 2011 Measures for the Management of Sand Excavation in River Courses in the GZAR was only 20,000 yuan. This was a low amount compared to the high profits to be made in selling sand, and it is no wonder that such fines did not discourage illegal operators from mining sand. The fine was increased to a maximum of 500,000 yuan in the GZAR in 2016, which is 25 times the old penalty. The provincial media called it a "heavy fist" intended to stop illegal sand mining (C. Li 2016). However, the outcome of the increase in fines is not clear yet because it is a recent law.

Some authorities fight for fines in order to enhance the income of their departments, sometimes resulting in disputes between two levels of government concerning illegal sand management. Lu (2017) reports on a fight between different levels of the police force in Shaanxi Province on December 31, 2016. The Law Enforcement Brigade and Public Security Bureau of Xingping City had detained vehicles and machinery belonging to illegal sand miners in Hu County and were in the process of escorting the vehicles and machinery to Xingping City. However, Hu County policemen and county officials of related departments blocked the road in front of the prefectural police and halted the Law Enforcement Brigade of Xingping City. The Hu County government was said to be protecting sand-mining sites, while the city government was attempting to punish the illegal sand miners. Witnesses said that a dozen police officers pushed and confronted one another, and three gunshots were heard at the scene (Lu 2017). A similar situation reportedly took place on the Yangtze River (see Wu 2008).

Mr. Huang, who was "stealing" sand in the Maowei Ocean with his team, also experienced the complex fine system. After he refused to pay a bribe to local officials, his dredger was confiscated by the joint law-enforcement teams of the Water Resource Bureau, the Maritime Bureau, and the Marine Supervision Bureau. He was told that each of the bureaus would fine him 20,000 yuan if he wanted his dredger back from the Qinzhou Port. He then went to the office of the Maritime Bureau and was told, "You need to pay in total 110,000 yuan; the Land Bureau also asks for a fine. There are five bureaus, not just four." Mr. Huang thought that the fine was too high and he left the office without paying in October 2018. In December 2018, he paid 99,000 yuan to take back his dredger. After two months of payments, he obtained his dredger ("The boat is rusty," he told me) but also received an invoice for the confiscation of illegal sand-mining revenue on February 1, 2019 (Figure 30). He regretted that he had not paid the initial bribe of 3,000 yuan that the local officials originally asked for. If he had paid then, his dredger would not have been expropriated, as he would have been notified in advance of the arrival of the joint law-enforcement team. In other words, a multitude of fines can be avoided through corruption. According to Mr. Huang, in July 2019, the bribe for one ship was 4,500 yuan in Huangwutun Town.

Indeed, governmental corruption is a pervasive issue in China, and it weakens the formal institutions in the region. "Money can handle anything in China. If you have money, then you do not need any license, because money is the most powerful license," a businessman told me. Corruption has therefore always been part of river-sand management. Seemingly respectable

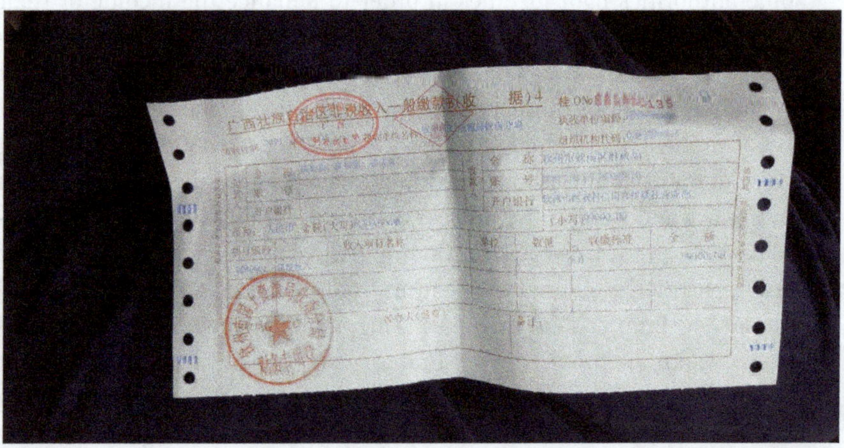

FIGURE 30 An invoice for the confiscation of illegal sand-mining revenue
SOURCE: FIELD DATA, MARCH 2019

government officials have been involved in illegal sand extraction and sand overexploitation in China.

Government officials do not deny their connections to illegal sand miners. During my field observations in 2017, a prefectural government official told me that if I went out for a beer with him, he would call his "friend" to meet me. This friend, who originally came from Hubei Province, was the major player in the sand business in Qinzhou City. The official said: "If I just call him and tell him to talk to you, you might have a chance to talk to him. Otherwise, he will definitely not talk to you or even look at you" (April 6, 2017, Qinzhou City). I was shocked by the words of this high-ranking official. His close ties with a prominent businessman could not just be a simple matter of friendship. It was all the more shocking when the official repeatedly told me that sand mining was in all cases linked with gang activities, yet he was apparently not ashamed of his connection to criminals.

Cases of corruption in Qinzhou City were also reported in the media. One such case took place after an auction in the Qinbei District of Qinzhou City in autumn 2011 (Jiang 2015). Mr. Shi's mining company received a license for two different river sections: the area from the electric pumping station in Nameng Township to the Niupi power station in Dasi on the Maoling River, and the area from Nalu in Qingtang Town to the Dingmeng section in Jiulong Township.

However, Mr. Shi began mining before he had actually obtained his license. He said that the license was only valid for one year. He had paid the entire fee for one of the two sections but still owed part of the fee for the second section. Consequently, the Department of Water Resources had not issued Mr. Shi a sand-mining license within the predetermined time period. In order to begin harvesting sand as soon as possible, Mr. Shi repeatedly went to the Department of Water Resources to ask for the relevant documents and certification, but he was denied and thus decided to start mining without a license (interview with a cadre of Qinbei District Government, March 2015).

A more serious obstacle to Mr. Shi's business was posed by the many illegal sand-mining activities occurring in his two river segments. Mr. Shi repeatedly reported the chaotic situation to the Department of Water Resources of the Qinbei District, asking the authorities to strengthen their efforts against illegal sand excavation and to protect his profits. However, the director of the department said that the department lacked funds to speed up the accreditation process and move in on unlicensed sand operators. In order to acquire his sand-mining license and combat the illegal sand-mining companies, Mr. Shi gave 200,000 yuan to the director, Mr. Lao, and requested him to take action against unlicensed sand miners. Mr. Lao accepted the money. The bribe was discovered by the Commission for Discipline Inspection in 2015. The District

Procuratorate prosecuted Mr. Shi for bribing Mr. Lao. Mr. Shi was sentenced to one year's imprisonment, which was suspended for two years. Later on, Mr. Lao, the former director as well as the party secretary of the Department of Qinbei Water Resources, was found guilty of accepting bribes, and he was sentenced to eight years in prison (Court of Qinbei District 2015).

Between 2013 and 2014, all of the directors of the Water Resource Departments of two districts and two counties were arrested in Qinzhou City on charges of corruption. These directors and about 40 other officials were particularly notorious in Qinzhou City for corruption and the mismanagement of a water conservancy project (Wang and Xiong 2015). It was an open secret that even a regular worker of the Water Resources Bureau who had a certain amount of power could become involved in sand mining and benefit from this activity.

In 2005, the Guangdong government announced that Guangdong Province was the first province to implement the regional law on sand management. This law was intended to eliminate corruption in the process of issuing a sand-mining license. The case of illegal sand mining by four companies indicates the severity of corruption between 2012 and 2013. According to an article in Nan Fang Daily (Zhao 2014), the four companies united in order to control the sand-mining industry and maintain their huge profits from sand mining in the West River—the third-longest river in China, and the longest river in South China—around Yunfu City and Zhaoqing City in Guangdong Province.[6] The conglomerate could decide the market price of sand, and it controlled sand-mining business activities in the West River between 2011 and 2012. To achieve this, the conglomerate paid bribes to officials in five counties and two prefectures, some of whom held power as high as the provincial level (Zhao 2014). The People's Procuratorate of Guangdong Province initiated investigations of 30 government officials after shutting down the illegal sand company. Under the protection of those officials, the joint company grossed more than 5 million yuan each day by illegally mining and selling sand. Over a period of 11 months, the conglomerate scooped out 24.18 million cubic meters of river sand and made a profit of 580 million yuan (Zhao 2014).

The case reveals the loopholes in formal sand-mining management institutions and the problems with implementing the license system in Guangdong Province. The director of the Department of Water Resources of Guangdong Province, Huang Boqing, who was sentenced to 15 years in prison for accepting bribes and fined 5 million yuan in 2018, had pointed out the difficulties and proposed a solution several years earlier:

6 The West River also passes through Guangxi; it is the fourth-longest river in China in terms of its total drainage area.

> Practice has proven that the application of licensing methods cannot objectively reflect the value of river sand as a scarce resource. The sand-mining field may become a hotbed of corruption in particular areas. Therefore, now we will implement a fair and equitable bidding method. The implementation should be called a "permit under the sun" (*yangguang xuke* 阳光许可, meaning a transparent permit).
>
> SUO, XI, and ZHANG 2012

One perspective on formal sand management in China is, therefore, that it is based on a "collusion between power and money" (Zhou and Ouyang 2008, 59), and that corruption will still be an important issue in formal sand management in the future. This collusion hampers the effectiveness of national laws and regional institutions, damages state resources, and harms national and public interests. "Collusion" takes different forms, but most of the time, the officials on the inside know clearly which boats, sand-mining sites, or selling sites have a special "background," or a hidden political power sustaining them. This kind of informal relation has weakened or destroyed formal authority and institutions on the prefectural level in many areas.

6 The (Non-)Governance of River Sand at the Township Level: A Lack of Authority

The introduction of the sand-mining licensing system did not create an effective form of sand management at the prefectural level. What was the effect of the implementation of sand-mining institutions and related laws at the township level? Did they help the local government to be more effective in sand governance? What did sand management on the township level look like in practice? These were questions that I posed during a group interview with township officials of Huangwutun Town on April 7, 2017.

The discussion revealed that the officials generally lacked the authority to govern river-sand mining effectively. The officials discussed the useless laws, fragmented institutions, and rare involvement of a superior government in the effective governance of sand-mining issues at the township level. The vice secretary of the CCP of the Huangwutun government stated:

> We did not get any help from laws [in managing sand issues]; what we have are duties and responsibilities. According to the principle of territorial management of the Chinese government, we have the direct duties of sustaining stability, guaranteeing safe production, and supervising and

> managing our administrative territory. The institutions have an impact on the monitoring system; the township government does not have much administrative power or lacks the relevant management power to fulfil the duties that we have. Consequently, we have to try very hard to prevent and avoid accidents and troubles, and any kind of social unrest. Sustaining social stability has become a major target in terms of illegal river-sand mining.

Without the help of laws relating to river-sand mining on the township level, not to mention the problematic management on the level of the prefectural government sketched in the previous section, illegal river-sand mining is common at the township level. The director of the Construction Bureau, nicknamed Sange, pointed out that there are 91 sand-selling sites in Huangwutun. None of them possessed a license, nor did they pay any fees or dues to the township government. Furthermore, one deputy mayor pointed out that no other authority (neither the county nor city government) had gathered revenue from the 91 illegal sand miners or businessmen since 2010. The local officials were aware that they lacked the authority to obtain fees and revenue. On the other hand, the officials were criticized by the villagers when they pointed to this lack of authority as a reason for not intervening in sand-mining.

Sange recalled in the same group interview that there had been massive government campaigns between 2008 and 2010. At the time, many authorities from city and county government levels cooperated with the township government in order to perform remediation measures in the aftermath of illegal sand mining. The township government participated in the two government campaigns as an assistance unit but not as a power unit. "They [illegal sand miners] stopped for several days," Sange said. Yet,

> significantly, the lack of follow-up management measures of the government damaged the previous governance efforts, simply because the campaigns were just short-term or one-time actions. The rare presence of the county authority and the rare conduct of joint campaigns contribute to the insufficient river-sand management. This means that there is no long-term effective approach [for monitoring illegal miners]. In my view, the government cannot establish long-term governance mechanisms and an improved management situation.

The long-standing activity of many of the illegal miners, the presence of overloaded trucks, an illegal transfer system for transporting river sand, and rising social conflicts should together provide sufficient evidence to demonstrate the problems of either the implementation of the sand-mining system and related laws, or the inherent weaknesses of the institutions. Indeed, a deputy mayor made a clear statement: "Before 2009, there were nine departments engaged in issuing sand-mining licenses at the local level. After 2009, there is not even a single department that dared to issue a sand-mining license."

It seems that no local sand miners can obtain a license, and they are therefore considered illegal. They cannot participate in the auctions at the prefectural level because they cannot afford to fulfil the basic management requirements. Moreover, to get a license, one has to obtain permission from each of the nine departments involved. Each department, in turn, needs a month to give this permission — "to affix a stamp." It is therefore extremely difficult to obtain a license.

In addition, the deputy mayor mentioned during the group interview that different township authorities possess power in the sand-licensing system and have the authority to implement the supposedly standardized regulations. The vice secretary of the township CCP added:

> Sand-mining management has become a management field common to various departments; owing to problematic management circumstances, the common management means no management. This has led to a confusing bureaucracy, in that each department has its own responsibility. However, no one really has full responsibility. As a result, no department wants to govern sand mining.

In other words, when citizens make a complaint about illegal sand-mining, local authorities are often unable to act due to the involvement of other authorities on several levels. Moreover, the "collusion between money and power" discussed in the previous section also plays a role on the township level. In short, the reasons why the issuing of sand-mining licenses has failed are manifold.

The vice secretary further argued that empowering the township government—within the current top-down government structure—is the key solution to counteracting illegal sand mining. He argued that:

> It is the local power that might be able to govern sand-mining activities. If we had the authority to govern sand, that would be another situation (much better). We would build up a standardized management system, for instance, to establish a sand cooperative company; all sand businessmen would have to join the cooperative company, and the organization would democratically elect their leadership. In this regard, some of the biggest sand-business people would be chosen to be the leaders.

> A self-governing structure of sand miners and businessmen would be formed. In so doing, they would establish an inner governance structure among all the sand dredgers or dealers. The committee could, for example, collect a garbage fee or even revenue. This approach could benefit the government by raising our income and reducing trouble somewhat. Unfortunately, the government is losing state resources and facing many cases of social unrest due to illegal sand mining, rather than benefiting from it.

During the discussion, the vice secretary emphasized that in his opinion sand management can only be achieved through dependence on local power, as the governments above the township level clearly cannot handle the issue. He explained that "local power" not only includes the local government, but also the sand-business people and sand miners, and it thus implies the power to govern sand-mining issues as well. He therefore proposed co-management between the township government and the committee of sand companies (excluding the villages and villagers) as the best solution.

But co-management between the township and the sand companies was never implemented; rather, authority on the local level was centralized with the introduction of the River Chief System in 2016. After the implementation of the River Chief System in Qinzhou City started in August 2017 (four months after my group interviews in Huangwutun Town), the prefectural and county governments extended their governance and monitoring down to the administrative village level. As mentioned previously, in the beginning of 2018 all the illegal sand-mining sites in Huangwutun Town were closed down. In addition, Huangwutun Town's section of the Maoling River was auctioned to Sanxin Company, which is run by four businessmen; this company paid 26 million for the right to mine there. This time, the government did not issue a sand-mining license, however, but issued permits for "dredging engineering in navigable waters." There are about 40 individual miners dredging in the name of the company (interview with Mr. Huang, July 2018). Meanwhile, other sand miners without permits have been forced to stop their mining activities. "Dredgers cannot drive to harvest sand; trucks cannot drive to load sand," stated the sand miners. Mr. Huang confirmed that the government's "comprehensive banning of pumping activities" forced him to shift to another remote town to continue his sand-mining activities in March 2018.

It seems that sand mining was going on under another name. I asked my sister Ke Zhu to check the river in Naxia Village in September 2018. She found that there was a big dredger in the middle of the river and that the water of the river was still polluted. The River Chief System had failed, despite the appointment of multiple river chiefs to manage the river and sand mining. They did not protect the water resources but redefined sand mining as "dredging engineering" in order to allow sand extraction to continue, due to the consistent demand for river sand. As I discussed previously, in March 2019, the Huangwutun Town government reported that 75 illegal sand-mining sites had been demolished by direct order of the Central Environmental Inspector. Eventually, it was pressure from the Central Environmental Inspector that ended the mining in the Maoling River in Huangwutun Town. For instance, the Sanxin Company was forced to stop mining, even though it had paid a fee and received permission to mine from the Qinnan District Government. Nonetheless, illegal sand mining continued: Mr. Huang was still mining sand in another town of Qinzhou City in July 2019.

7 Discussion and Conclusion

In this chapter I have described how river-sand management in China is handled by various levels of government, and I have explored the establishment of new institutions and laws, such as the sand-mining license system and the River Chief System. My intention is to provide a better understanding of the environmental crisis and resource scarcity due to China's rapid economic transformation. I aim to contribute to debates within the field of political ecology about China's government structure with regard to natural resource management (Wang 2013, 2018; Kostka and Hobbs 2013; Xie 2015; Yeh 2015). Yonghuan Ma et al. (2018, 54) state: "Environmental pollution and ecological degradation seriously affect each other. ... The emergence of these problems is related to the lack of top-level planning, blind development and excessive exploitation." They argue: "Institutional guarantees, legal guarantees, supervision guarantees, and service guarantees are the four magic weapons in the natural resources management system" (57). This chapter has paid attention to the different levels of government and the power dynamic across these levels.

In my view, political ecology can overcome the shortage noted by A. Wang (2013, 369) when he argues for studying the implications of the environmental cadre evaluation system for environmental regulation and governance in China: "Little academic work has attempted to develop a broader theory of the functional relationship between bureaucratic mandates and Chinese law. For the most part, researchers have studied the two governance systems in isolation—(almost) never the twain shall meet." Political ecology includes both formal institutions, including laws and policies, and informal institutions, such as religions and customs, to explore crucial aspects of the human–environment relationship. In the case of river sand governance, this approach can include both different levels of government (vertical politics) as well as organizations within the same level of government (horizontal politics). The interaction between laws and government authorities, and how national policy and regional laws engage with local river-sand mining, are explored in this chapter.

By providing a case study of river-sand management, I highlight the interactions between formal institutions on various scales within China's top-down authoritarian system. The contradictory laws and the fragmentation of authorities between and within different levels of government have led to a diversity of discourses and varying perspectives, increasing rather than decreasing the number of stakeholders and conflicting interests when it comes to sand management. The overlapping laws reflect that river sand is not perceived to be a resource on its own but has been considered to be part of various other resources. Although Decree No. 320, issued by the central authorities in 2002, appointed the Department of Water Resources at various governmental levels as the main responsible authority, this department cannot effectively manage river sand or protect public interests on its own due to the involvement of numerous other departments and offices.

Moreover, the power vested in the Department of Water Resources to issue sand-mining licenses and perform "on-site supervision" of sand mining has become a way to generate profit both for illegal mining companies and for the officials of the department at or above the county level, where corruption occurs. The failure to achieve sustainable management of river sand in Qinzhou City in the last two decades reflects the ineffectiveness of the river-sand license system that was introduced in 2002.

Therefore, the central government in 2016 initiated a political reform, the River Chief System, to deal with the issue of fragmented authority and the inability to hold government officials responsible for effective management of river resources. The reform has forced governments at various levels to put great effort into eliminating illicit river-sand mining by taking a close look at the levels of power involved in river-sand governance. Within the power structure of authoritarian China, government leaders at all levels have the strongest power to execute laws in their respective political territories. They, as "river chiefs," have thus been made responsible for the coordination of resource management activities in their jurisdictions.

The River Chief System is based on the centralization of power and represents great ambition on the part of the current central government. However, the effectiveness of this reform can be questioned on the basis of my research findings. The following paragraphs present four criticisms of the River Chief System reform.

First, this reform is top-down, reaching from the central government to the county level (and down to the administrative village level in the GZAR). As a result, the role of the CCP has been strengthened, for next to the governor, the highest CCP chairman on each level is the one responsible for river and lake management and water conservation. This reform accomplishes the recentralization of authority (see also Kostka and Nahm 2017) by stressing the leader's duty as well as the assessment of the effectiveness of regional officials, whose political careers depend on their success or failure in river management. The central government emphasized that political reform would be the paramount management subject with regard to sand governance, which increases the significant role of the party-state. Chien and Hong (2017, 58) explain: "River leadership is assigned to certain prefecture-level cadres, whose career advancement depends on achieving specific goals related to the quality of rivers for which they are made accountable." They argue that the personal assessment of major officials makes the system effective.

My findings demonstrate that it is not the local river chief who effectively manages river sand in Huangwutun Town. The 75 illegal sand-mining sites there were removed by the central environmental protection inspection team, which in effect has the far-reaching power to stop illegal sand mining and other polluting activities. Gilley (2012) argues that China's party-nation regime and system of authoritarian environmentalism is based on bureaucratic policies and leaders' responsibility. Both are more important and more effective than laws in resource conservation and environmental protection in China, as also demonstrated in Huangwutun Town, where illegal sand mining has continued after the implementation of the River Chief System in September 2017. Yet, even the effectiveness of the central environment protection team is in doubt, since illegal sand mining has persisted in Qinzhou City. As Li and Shapiro (2020, 55) conclude: "In the end, environmental enforcement becomes contingent on the arbitrary dictates of the central inspection team and the fears and ambitions of grassroots cadres, as opposed to transparent, accountable environmental laws and standards." Wang (2013, 435) argues:

> Top-down administrative monitoring and market measures typically perform poorly in environments with low levels of public accountability, a tendency toward capture, and structural incentives for collusion. Top-down monitoring and market measures do not by themselves fundamentally alter these negative influences on environmental cadre evaluation performance.

Second, the River Chief System was an administrative reform, grounded in policy rather than law. There is an ongoing debate on the respective role of law and of executive management when it comes to China's environmental crisis. The central government prefers administrative-based measures, highlighting the role of the party-government in controlling sustainable resource use and environmental management. Significantly, the recently enacted regional laws in Guangxi Province and the new national political reform reveal that the issue of river sand has become a force by which the entire river-health management system may shape politics in modern China. It seems that, despite facing a series of failures in sand mining, the government still prefers to prioritize bureaucratic action rather than laws. Laws are still largely ignored by the latest political reform in terms of river conservation and sand management, even though there is some law-based management at the provincial level.

Third, I argued that it is too simple to perceive the Chinese government structure as a dichotomy (top-down, central vs. local, or vertical vs. horizontal). China's crisis in river-sand management is not a consequence of the central government's top-down approach, but of its failure to recognize the various levels of government and numerous horizontal departments involved. National environmental policies are diluted by overlapping laws, ambiguous ownership, fragmented authorities, and lack of public participation in the boom of the sand industry.

Fourth, the new reform excludes indigenous people from participating in environmental decision-making, even though it relates directly to their land, resources, environment, and culture. This is a major characteristic of authoritarian environmentalism, which Gilley (2012, 287) defines as "a non-participatory approach to public policy-making and implementation in the face of severe environmental challenges." The formal institution is currently transforming, but the role of villages and/or villagers—the direct stakeholders—is excluded, marginalized, or undermined by decision-makers. In particular, the governance system has excluded the local villages that have played a vital role in rural common-pool resource management and benefit sharing. Thus, indigenous resistance at the village level against state-driven sand mining and market-driven sand mining has been reported. This has resulted in rising tension among river-sand miners, local riparian minority communities, and government agencies, which will be discussed in the following chapters.

Indeed, river-sand mining management is not an isolated activity but is embedded in China's political, economic, institutional, and cultural context. Apart from government perspectives on river-sand ownership and management, other views on property rights related to river sand surfaced during the analysis of discourses on its ownership. The different discourses about the ownership of river sand have played a crucial role in the commercial appropriation of river sand, as well as the related management practices and conflicts.

Indeed, the property-rights issue and the relevant laws greatly impact sand management, which has shaped the sand-mining reality on the ground; however, the question of who should wield these property rights, and what form an effective river-sand management system might take, remains debatable, as illustrated in the next chapter, which describes the varying perspectives on property rights and river sand in the Zhuang villages.

CHAPTER 6

Zhuang Villages' Perspectives on Property Rights

"This is Our Sand"

As I have shown in the previous three chapters, there are several formal institutions, such as laws, government agencies, and administrative village committees, that impact river-sand mining activities. I have also explained the governance roles of lineage organizations and worship customs at the level of natural villages in the introductory chapter and chapter 2. In this chapter, I take a closer look at how sand mining is negotiated at the level of the natural village between different actors and levels of rural village governance, and how sand is slowly becoming commercialized instead of being considered part of the villages' common resources.

This chapter focuses on river-sand mining from the perspective of the Zhuang villages. The ownership debate and the crisis around river-sand mining are explored through case studies drawn from the six Zhuang villages that were selected for this study. Significantly, there is no unanimous viewpoint on river-sand mining in any given Zhuang village. The perspectives of actors are rather diverse and contradictory. Nevertheless, it is clear that in practice the majority of the natural villages in this study have negotiated a sand deal with illegal miners either in their own name or in the name of their production team, rather than in the name of the administrative village government.

To address the circumstances of river-sand governance from the perspective of the villages, this chapter asks the following questions: What sand are the villagers actually talking about? Who is involved in making sand-related decisions and in management? What solutions do the villagers end up with? And what institutions or laws are applied in the rural sand-mining context and invoked by different villagers?

John Loughlin (2018, 197) argues:

> The primary meaning of territory is to denote a physical space, marked by recognizable boundaries within which societies exercise control. For both humans and animals, this control is connected to food gathering and providing security for themselves and their offspring against competitors, whether from their own species or others.

 | DOI:10.1163/9789004505919_007

In China's system of property rights, communities have usufruct rights to those parts of land that are not part of a homestead and are not under individual cultivation, and they can use them as pastures or for harvesting useful products (Ho 2015). Villagers consider river sand as part of land and territory, legally belonging to the collectively owned land and controlled by the usufruct right system. When asked about sand, the villagers stated: "This is our sand." This claim was supported by some officials of the administrative village and township governments. It is necessary to point out that this belief is also related to the clearly-bounded territory of the community of local Zhuang in rural Guangxi.

As Ulloa (2015, 321) points out:

> When the dynamics of environmental disputes are analyzed, the claims and resistances, as well as proposed development alternatives, involve other views about nature and spatiality, since they are not only about access, control and effects, but about other ways of being and living in a specific territory.

This chapter demonstrates that the concept of territory is important for Zhuang rural villages in terms of identifying ownership of land, establishing a right to use land and other resources, claiming membership and belonging, providing security through social relations, raising resistance against wrongdoers, and simply for determining their ways of being and living. The chapter begins by discussing those who take the sand from the villages' territories without local permission, who are considered thieves (*zei* 贼). The use of the term "thief" shows that from the perspective of the villagers, sand issues are part of "territorial rights and justice" (Nine 2018, 208).

1 The Emergence of Sand "Thieves" and Ownership Problems

"Thief" is a negative term used by local villagers and officials to refer to a person who steals public sand from the river to sell. The so-called thieves are those insiders and outsiders who dredge river sand for the market and benefit themselves. Sand thieves in Mei Village first emerged in the early 2000s when the price of river sand started to increase. In 2007, Pan Jiu, who was then 40 years of age and a resident of Nayang Village, was the first such "thief" in Mei Village's river. He first brought dredgers and appropriated sand from streambeds of the Maoling River. Soon after, Pan Jiu began dredging in the river section near Mei and Na Villages, which resulted in rising villager opposition: "The Nayang thief is stealing our sand!"

Once the outsiders, such as Pan Jiu, could not be kept away, since the government did not stop unlicensed sand mining, local villagers also began to mine sand for the market in Mei Village. The new village head, Mr. Fufang, who had been elected in 2011 after the old village head died in 2010, explained:

> In 2008, there were three young men (around 27 years old) in our village, belonging to three different lineage branches, who began the sand business. They were Mr. Fuben, Mr. Fubao, and Mr. Fuhan. Soon, Mr. Ge joined them. Those men had previously gone to the cities to find jobs. However, they did not like the menial work or the low wages, such as on construction sites or in factories. Then they realized there was a huge demand for river sand for building houses and roads. Meanwhile, they had seen how villagers from other villages did sand excavation in rivers and how they profited. They believed that a sand business would be a good choice for making money, as it offers a big return with little initial investment. At that time, they were afraid that the other villagers would oppose their business. Therefore, Mr. Fuben and Mr. Fubao enlisted the grandson of the old village head, known as Mr. Fuhan. It was hoped that since the three of them came from diverse lineage branches, it would reduce the opposition against them.
>
> interview in Mei Village, October 24, 2014

In terms of common-pool resources management in general, Ostrom (1990, 36) argued: "In every group there will be individuals who will ignore norms and act opportunistically when given a chance. There are also situations in which the potential benefits will be so high that even strongly committed individuals will break norms." This was the case in Mei Village when the three village residents launched their sand-mining business. I was told later that the news of the three young men pumping sand from the river spread quickly throughout the village. Discussions about this topic were tense, especially because several elders emphasized that river sand is communal property. This point of view was highlighted repeatedly by some elders and other local residents: "Sand extraction for personal use or for house construction is permitted, while selling sand for personal profit is prohibited because it will lead to sand overexploitation and damage to the riverbank, land, and environment" (interview with Mr. Yaozu in Mei Village, October 30, 2014).

In contrast, the three young men argued that if they did not harvest the sand for business, outsiders from other villages would pump out the sand, since the elders had not found a working solution for the protection and prevention of outsider harvesting in Mei Village. In addition, one of the young men claimed, "Sand will come again with the seasonal floods." The elders responded: "An empty river without sand—riverbanks, land, crops, trees will collapse when the flood comes. These things are happening. Our land is disappearing." Some women complained that the deepening of the river channel made it difficult for them to farm on the other side of the river. "It is too dangerous to cross the river by bamboo raft. I have not grown any crop on my field for three years," one female elder stated (December 20, 2014). Her remark indicates the existence of "different ecological ontologies" (Horowitz 2015, 241) among different generations in Mei Village. The debate reflects the ambiguity of ownership and institutions (e.g., use rights of natural resources) regarding river sand for the villagers. Interviews with the sand miners show that the pressure from the villages is substantial. There was talk of "throwing stones at the thief" when the elders organized people to stop the "thief" who was dredging the riverbed (this is discussed more elaborately in chapter 7).

In early 2010, Pan Jiu's dredger was confiscated by the people of Na Village because he was illegally mining river sand in their village's territory. "Pan Jiu must pay 6,000 yuan if he wants his dredger back," the villagers of Na Village demanded. Pan Jiu did not want to pay any money. He therefore called the police of Xintang Town, who came and investigated the situation. Pan Jiu asked the police to force the people of Na Village to return his dredger. On their side, the village residents argued that the police must stop the illegal mining that had damaged their land and left them bereft of their river sand. They also said that the police should punish the "thieves" illegally mining the river. The police could not handle the situation, however, and simply reported it to the higher-authority agency (the Water Resources Bureau at the district level). During this period, the people refused to return Pan Jiu's dredger. Eventually, in order to continue mining sand to supply the market demand, Pan Jiu paid the unofficial fine imposed by the village to get his dredger back. Na Village was successful in this case because it is a big village with more than 1,000 residents, and because its elders managed to unite many villagers to ask for protection for their resource. However, not all villages were as successful as Na Village in uniting their people against sand thieves.

2 The "Blame Game" and the Complexity of Rights on the Commons

Sand-extraction practices have created much confusion and led to many "blame games" in the villages that have access to the river. Confusion begins with the very definition of sand as a communal (or common) resource, and the ambiguity that comes with the rights of exclusion and inclusion in the appropriation of sand. Moreover, the plots of land adjacent to the river as well as the riverbanks are held in the form of "village common" property or "private" property, which has legitimacy in the Zhuang villages; these rights are handed down from generation to generation to the point that they become almost uncontested. Just as in many other contexts around the world, these property rights are rarely written down or formalized, and they only have legitimacy at the local level.

The failure of the local and national governments to clarify the ownership status of water resources such as sand compounds the confusion around access to and use of sand, as well as the extraction of the resource for commercial purposes. This frequently leads to finger-pointing, as villagers blame people (both insiders and outsiders) for harvesting sand illegally without necessarily defining why their activities are illegal.

To elicit responses to questions around resource ownership on the local level, I conducted a focus-group discussion on April 10, 2017 with ten respondents. We addressed some of the questions I asked in the introductory chapter, including the following: Who is monitoring the public river sand? Who is sanctioning the offenders? To whom are such cases reported, and what action should be taken against the disobedient persons? How do socioeconomic disparities play out in sand management? The discussion showed that local villagers and illegal miners are somewhat overlapping groups. The abovementioned miner Pan Jiu is on bad terms with the other villagers in his village, Nayang, because they see him as an illegal sand miner and a thief. They blame him for disregarding the only bridge that connected both sides of the upstream part of the Maoling River, which was built in the 1960s. The bridge was crucial infrastructure that connected the village to the outside world, yet it almost collapsed due to the heavy extraction of sand and related problems, such as flooding.

Even though the name of Pan Jiu resonated widely in other villages, such as Mei Village and Na Village, it was seldom formally mentioned by participants in the focus group and survey in his own village of Nayang. Although the participants did not name Pan Jiu, everyone knew that he was the subject of the discussion since he was the only person still mining sand in 2017. None of them wanted to cause any trouble since Pan Jiu was reputed to be unafraid of any opposition, and they knew that he was not afraid to use violence, either.

The villagers stated that "he" began the illegal activities in 2007. Immediately, villagers organized a meeting to discuss a possible sanction against his actions. However, being members of the same lineage as Pan Jiu, they could not reach a consensus; instead, intra-group conflicts and disputes occurred. The problem was that the villagers lacked the ability to enforce their rules or penalties on each other because they were all from the same lineage. Indeed, all the male villagers have the same family name, and for generations, brotherhood has been very important for this village. It has helped them to achieve collective goals, such as building a bridge and a village road. Yet in this case there was a conflict between how the villagers value their resources, such as their bridge, river sand, and environment, and how they value the members of their community. Due to the lack of effective sand control and collective action by the village and the government, Pan Jiu continued extracting sand, eventually damaging not only the old bridge (built in the 1960s; Figure 31, a), but also the new bridge (built in 2012; Figure 31, b).

Ironically, this new bridge was the product of collective action by Nayang Village and beyond. The bridge had cost about 500,000 yuan, a considerable part of which was raised by Nayang Village and surrounding villages. Villagers had also contributed their labor to the construction of roads connecting the bridge. With the increasing number of sand dredgers near the new bridge and the possibility that it might collapse following the erosion of the riverbanks, villagers were increasingly concerned and called for a meeting with all sand harvesters who operated their machines near the bridge. No sand operators showed up at the meeting. Interestingly, information on the meeting disseminated through posters—no one was courageous enough to inform Pan Jiu or the other sand dredgers in person. There was fear that these illegal operators might cause harm to anyone disturbing their business and that personal relationships between those bringing the message and the operators would suffer. This tension could even spread to the messengers' families.

Significantly, the absence of formal sand governance and the lack of protection for public infrastructure and private riverbank land were highlighted by the participants in the focus group. When I asked the villagers about the party cadres' sand-governing activities, the only answer I received was, "We don't know; one cadre of the administrative village is living in our village, but he did nothing about sand mining." The elders hoped the cadres would do their duty and take action to stop the illegal mining. If the elders were to take action instead of the cadres, their deeds might result in a loss of harmony inside the lineage (which occurred in Mei Village; see chapter 7). The villagers greatly value harmony in their community, as expressed through collective worship, festivals, and honoring the land of the god. In their view, the village had developed solidarity in order to deal with environmental crises, such as drought and flooding, and to construct the public infrastructure during its long history, including dams, canals, flood-prevention levees, a road, the bridge, and

FIGURE 31 Pan Jiu's dredgers in front of the old (a) and new bridges on the upper Maoling River (b)
SOURCE: FIELD DATA, 2017

schools. The villagers of Nayang Village are very proud of their solidarity and harmony. Whereas the residents of Mei Village debated whether a household should contribute its land for public use (see section 3.2.1), the households in Nayang Village contributed their land, money, and labor to public projects. Residents of other villages and the cadres of the village committee confirmed the great harmony in Nayang Village, which has a very good reputation. This reputation may even help the village obtain more government projects through the project system, and it explains the reluctance to act against illegal miners from inside the village.

At what point does a villager become an illegal miner? Does it happen when someone's share of a commonly held resource surpasses that of other members of the commons, or when someone's activities are deemed to bring about serious consequences? An elder said:

> People here are quite afraid of them [illegal sand miners]. … But one day I could no longer bear their selfish behavior. I went to the sand-selling sites and warned them that they would have to rebuild the bridge if their activities were to eventually lead to its collapse. However, this did not concern them. He [i.e., Pan Jiu] has run his sand business for more than ten years without any license from the government or permission from his fellow villagers.
>
> focus group, Nayang Village, April 10, 2017

Pan Jiu and some of the other insatiable "illegal" sand miners are considered members of the commons. They are blamed for the possible collapse of the bridge and for commercial exploitation of sand because their actions are linked to what villagers consider selfish extraction, impinging upon the needs of others. Thus, the term "illegal" is only used in contexts where actors seem to pursue selfish gains by exploiting commonly held land or someone's private land. What clearly emerges in the narrative of the confrontation between the villagers and Pan Jiu is an indication of the complexity of rights. Rural villagers can mine certain amounts of river sand for free for personal use, for constructing a house, or for growing Chinese yams, while the village can mine river sand for public construction, such as building a new road, a channel, or a temple. The description also highlights the complex (socioeconomic) power relations involved in allocating rights over common resources. The problem is that other villagers, who would otherwise have access to free sand, are forced to buy it from Pan Jiu. This case shows that both elders and village cadres failed to sanction the violators and ended up tolerating long-term illegal sand extraction in the local community. Furthermore, it reveals how social norms and values such as communal harmony and peace influence issues of law and order: the villagers failed to take direct, proactive, collective action and make complaints to the government against a few wrongdoers in their community because they did not want to disturb the village's harmony. There also seems to be high regard for communality versus individuality, related to the Zhuang people's customs, beliefs, and sociocultural system.

In the following, I present a case in which township officials engaged in sand-mining operations in two other villages in the same administrative village, Nagan.

3 Selling Sand on the Instructions of Government Officials

Apart from disputes and conflicts inside villages and lineage groups, an intervention by township government officials in Xintang made the sand management crisis worse. When conflicts over river sand in Mei Village and Na Village were reported to the township government in 2008, officials did not show up to resolve the disputes. It was not until 2011, when the development project system, introduced in chapter 4, was implemented in Xintang Town that the government actively intervened through a method that became known as "killing three birds with one stone." The township officers practiced three strategies:

1) The officers encouraged the village elders and village head in both Mei Village and Na Village to sell the common sand and place the proceeds in a collective fund.
2) The officers encouraged the villages to use this fund to cooperate with the government in improving public infrastructure inside the villages as part of the project system.
3) The officers arranged things so that they or their relatives could buy the sand themselves. The mayor, the director of the justice bureau, and a vice-director of the traffic bureau of Xintang were all involved in the signing of the sand-mining agreement, and thus managed to buy sand at a relatively low price and derive a huge benefit from it.

The selling points for "killing three birds with one stone" were, first, that sand conflicts would be eliminated, since illegal miners would be excluded from the license auction, even if they could pay a higher price per cubic meter of sand; and second, that the village would collect the funds necessary to develop its infrastructure within the project system. However, the main objective of the arrangement seemed to be to deliver enormous benefits from this sand deal to the township officers themselves.

While the elders and the villages were hesitant, the sand businessmen, who were the nephews of the abovementioned officers, and the officers themselves visited the village heads of both Mei Village and its neighbor, Na Village, and asked the cadres of the village committee to help them convince the village heads to sell sand to them. The government officers told the two villages that they could divide the income from selling sand into two equal portions, rather than dividing the river sand itself according to physical boundaries. In other words, each of the villages would obtain half the cost of each cubic meter of sand extracted. This was exactly what they did. The price to extract a cubic meter of sand was 23 yuan, so each village could earn 11.5 yuan per cubic meter extracted. Meanwhile, the cost to purchase a cubic meter of river sand on the local market was 85 yuan in 2012. This means the miners earned 62 yuan as their profit. One of the miners in Mei Village said that his group offered to pay an extraction fee of 35 yuan per cubic meter, but the village head preferred the offers from officials because he wanted them to arrange the infrastructure projects mentioned above in the village (interview in Mei Village in December 2014). Without obtaining a license through an auction held by the Qinbei District government, the sand businessmen who were related to the township officers mined sand in the river sections of Mei and Na Villages between May 2012 and March 2013. Thus, they were considered illegal miners by the government of Qinbei District. The signed agreement with Mei Village and Na Village proved the guilt of the township officers. However, their relatives were considered legal miners by the local villagers.

During the selling season, both villages sent someone to calculate the number of loads of each dredger. These records were kept by the head of Mei Village for potential financial inspection, and they show that the highest sale in Mei Village between May 2012 and March 2013 was recorded on July 1, 2012, when the two dredgers operated by the relatives of the township officers extracted 2,077 cubic meters of sand. The total income from sand over the whole of 2012 was around 112,927 yuan in Mei Village, while the income of Na Village was 290,000 yuan because Na Village continued to sell river sand in another river section after the agreement ended.

The village head of Mei Village claimed that the government officials, the cadres of the village committee, and the village's party secretary knew very well how to convince the elders in both villages to participate in their scheme. The officials and other leaders argued that the villages needed new cement roads since more and more households had purchased cars. Furthermore, the improved infrastructure would reflect solidarity in the village as well as its modernity, which might increase the chance that the village's bachelors would find wives (about 20 single men, including the second son of the village head, had been seeking a wife in vain for several years at this stage). In addition, the authorities were aware that the elders were powerless to stop illegal miners, as the elders themselves acknowledged. Combining the customs of a rural village with the government's projects was thus a plausible scheme. The village head of Mei Village recounted:

> They [the director of the justice bureau and a vice-director of the traffic bureau] gave me an agreement and let me collect signatures and fingerprints from each household. I agreed with them because they promised to help us to construct "a new socialist village" (see section 3.2.1) for Mei Village, including a new ring road one kilometer long, a basketball court, a public toilet, and a pavilion. After the sand agreement was signed, a township official in charge of implementing the projects was sent to our village. I then have been working on these projects with him and even with the mayor, who invited the TV station and trucks to our village when we [the villagers] were asked to clean up the rubbish in the village. "The Construction of a Beautiful Village" was the name of this event. Then the event was reported on by the TV station of Qinzhou City.
>
> interview with the village head in Mei Village, January 20, 2015

As mentioned at the start of this section, it was the introduction of the project system that provided the township officials with this opportunity. As discussed in chapter 4, the "One Project, One Discussion" system stipulates that only the leaders of villages whose inhabitants are able to gather adequate funding can secure government assistance for the building of new infrastructure. Much of this funding comes from the sale of river sand, which has divided the village residents due to differing views on the advantages and disadvantages of large-scale sand extraction. For many villagers in Mei Village, the sale of river sand and the utilization of the proceeds did not produce the outcomes they had expected, leading to a growing mistrust of the leadership (i.e., the township officials and the village committee as well as the village head) and a breakdown of relationships within the village.

A further key observation here is that local resources were gradually becoming embedded in a larger capitalist market system due to the boom in the construction market and the actions of local officials (see also Chatterjee 1983). The local villagers had no option but to cooperate with the market. The following section, based on data from 2017, discusses similar cases in which three villages sold their sand, even without the active encouragement of township government officials.

4 Selling Sand to Defend a Village's Right to Common Resources

Nazhong Village and Naxia Village of Huangwutun Town also sold sand from certain river sections to gain an income from their common property. According to an interview with village officials, the popularity of "stealing" sand led to this phenomenon in their villages, eventually forcing them to sell river sand to secure their benefits and to exercise some level of control.

In early March 2017, cadres of Nazhong Village organized a meeting to discuss the sand crisis with the village's three production teams.[1] The members of two production teams believed that the only solution to reduce their losses was marketing the sand, while some members of the third production team pointed out that selling sand would damage the levees that were built in the 1960s to protect the fields from rising tides and seasonal flooding. The village production captain and the party secretary of the village committee then decided to sell the sand for 8 yuan per cubic meter.[2] Of this 8 yuan, 6 yuan were transferred to the village's common fund, while the remaining 2 yuan were distributed to any villagers who helped to count the quantity of sand that had been extracted that night.[3]

I arrived in the village on March 27, 2017, the seventh day on which the village was selling river sand. The production captain first thought I was a businesswoman who also wanted to do commercial sand extraction because I asked so many questions about the sand and the river. He told me that the profits from the section currently being mined were higher than he and his colleagues had expected. Satisfied with their gain, the cadres of the village were planning the commercial sale of sand from yet another section of their river.

The noise of the motors of the dredger could be heard in the village overnight. One evening I talked to the party secretary of the administrative village in his yard at around 7:30 p.m. He pointed in the direction from which the engine sounds were coming and said, "These are the sounds of the mining activities. We can hear them every night recently. It must be very crowded there since many villagers have all gathered together in order to share the income"

1 A production team was the basic accounting and farm production unit in the people's commune system in the People's Republic of China from 1958 to 1984; the term is still used in some rural areas, even though the household is now the unit officially responsible for the profits and losses of agriculture after the introduction of the household responsibility system in 1982. In these villages, people use "production team" to refer to the management of collective land and resources, such as forests, mountains, and rivers (see also Ho 2015).

2 A villager told me that most of the households signed an agreement to sell river sand during the meeting, but I did not see this document.

3 The dredgers can only mine during the night, when high tide arrives.

(Nazhong Village, March 28, 2017). He explained that sand selling was a common situation in riparian villages. Furthermore, more than a hundred villagers joined the sand-selling activities in Nazhong Village, from older inhabitants to children. They went to the riverbank when the dredgers began to extract sand in the evening and gathered in a place called Qiao Tou (桥头), which literally means "bridgehead" and is located on the side of the bridge where the sand-selling sites are located. Each night, there were at least three dredgers pumping out sand from the river close to Nazhong Village. And each night, the village sold at least 780 cubic meters of sand, which was valued at 6,240 yuan, and would distribute 1,560 yuan among the one hundred or so villagers present that night, while the village fund would receive 4,680 yuan.

The party secretary explained to me how villagers' participation in the selling activities was recorded and how the money was distributed. He said: "Two leaders of each production team would show up earlier in the evening and record who from his production team joined the activity during that night." He also explained that these team leaders had been chosen during the meeting on the sale of sand. The leaders of the three production teams also jointly calculated the quantity of sand that was quarried by the miners. The maximum number of dredgers the village's river section could hold was six. Notably, when the party secretary brought me to the riverbank, we met with an elder, a leader, the wife of a retired primary school teacher, and a group of young people who arrived in a new car. Members of the third production team also joined the group at the riverbank for a share, since the decision had been made. I observed that these villagers present did not engage in the mining itself. They were not even close to the dredgers that were scooping out sand in the middle of the river. "The villagers intend to protect the levees by jointly making sure that the dredgers do not extract the sand within the 15-meter prohibition line," an elder said (March 28, 2017). In fact, they were simply standing there, talking to each other, watching the quarrying process, and waiting to obtain their money when dredging was over. They had to wait for several hours each night in order to receive their money from the management groups. I was told that their "job" was to monitor the mining by standing around until they received a share (Figure 32 shows a temporary shelter used by monitors). The inclusion of all members has long been part of the public culture of the villages. Showing up expresses a sense of belonging in a village. This form of participation is based on the idea that each village member has an equal right to access public resources and to participate in decision-making and benefit-sharing. This kind of traditional popular culture in general has been observed by Siu in Nani Town in Guangdong

FIGURE 32 A shelter beside the river at Naxia Village from which sand mining was monitored
SOURCE: FIELD DATA, APRIL 2, 2017

Province, where she studied recycling rituals. Siu (2016, 97) states: "Through taking part in the rituals, the inhabitants sought affiliations that confirmed status and differentiated access to resources, claimed rights of settlement and tenure, forged political alliances, arbitrated conflicts, and secured social mobility." Sand mining in this context is similarly a ritual for the villagers, part of local popular culture, in which presence reflects membership in a communal group.

However, when looking at the larger picture of the local sand commodity chain, the way in which the villagers participate in the mining process shows how resource inequality and injustice emerge and increase in rural villages. The villagers were sharing the two-yuan benefit from selling the public sand—a trivial cost to the sand mining industry. But in the future, the villagers will suffer from the loss of the river sand, the damage to their land, the increased danger posed by the river, and the uncertainty of a potential environmental disaster. The villagers became the victims of the boom of overmining river sand while they were the least powerful stakeholders in this process. Even if they unite, there is little profit for them to share.

The cadres, such as the party secretary, village head, and others, did not receive a share of the profits. The village head told me the reason: "We dare not participate in the distribution of money, and generally do not appear at the sand-pumping site. We have been blamed for the loss of the river sand and cannot effectively govern the sand resource owned by the villagers. They [the villagers] are sensitive about whether we are corrupted or colluding with the illegal sand miners." However, the family members of these village cadres openly joined in with the sand mining during the night and took their share of the money. Even the parents of the party secretary were preparing to go to the riverbank on the evening I was interviewing their son.

Naxia Village, a village near Nazhong Village, similarly had to deal with the loss of river sand but had the additional problem that it had been without a village head for three years. On March 27, 2017, one of the village elders, Mr. Hu, organized a meeting on sand issues. I joined their meeting and conducted a focus-group discussion in order to explore the decision-making process for selling river sand. Four elders were present, with an average age of 75. Mr. Hu was the youngest at 70 years of age. He told me that a villager from Naxia Village had secretly sold the village's sand to sand businessmen. To reduce the loss of common resources, they decided to auction sand-mining rights to a 400-meter-long section of the river, with a starting price of 70,000 yuan. They would invite all the dredger owners and owners of sand quarries to attend the bidding, which was very similar to the government's auction.

In the meeting, I asked the elders who the sand belonged to. "Sand belongs to our village," three of them told me. Similarly, when I asked why they had not invited the village cadres to the meeting, they replied: "It is our sand, not the village administration's or the cadres'. The cadres have never visited our river and managed illegal sand extraction. It is really none of their business." In short, from the elders' point of view, sand belongs to the village and can only be governed by the villagers.

Mr. Hu added:

> Actually, river sand belongs to the state, but the state does not govern the sand either. Only in 2008 did they sell it [sand-mining rights] to the big companies. In order to protect the companies' benefits, the government sent extractors to destroy the illegal private sand-selling sites and banned the dredgers from appropriating sand from our river. After that, the government has not shown up, nor have the cadres of the administrative village, who sold sand before 2008 for 50,000 yuan. They only gave us 10,000 yuan for a water-conservation project. Now, we have to protect our sand

> and our dam. First, we will sell it and earn some common income, which is better than nothing. Second, we must protect our dam, since sand extraction is only allowed beyond 15 meters' distance from the riverbank.
>
> Naxia Village, March 28, 2017

In Naxia Village, it was the four elders who decided on the sale of the village sand, including the price and the way to distribute the profits. This is in contrast to Nazhong Village, where a mass meeting was held, with the party secretary, leaders of production teams, and many villagers present.

On April 2, 2017, Mr. Hu told me that Naxia Village had copied the style of sale that Nazhong Village had established, namely to sell sand at a price of 8 yuan per cubic meter. The village thus obtained 40,000 yuan in five days. Notably, I saw only a few elders and adult men involved in sand extraction in Naxia Village, while Nazhong Village's mining process included women, the elderly, and children. The absence of middle-aged men was a common situation in rural Guangxi during the year, with many men working in urban areas. The lack of men also weakened the capacity for resource management when facing wrongdoers and outsiders in rural villages.

During the sand-selling period in Naxia Village in April 2017, I conducted a survey among 29 villagers (including some who were not engaged in sand mining) on the ownership, benefit distribution, and governance of river-sand mining and related disputes. Figure 33 shows that it was unclear to the villagers

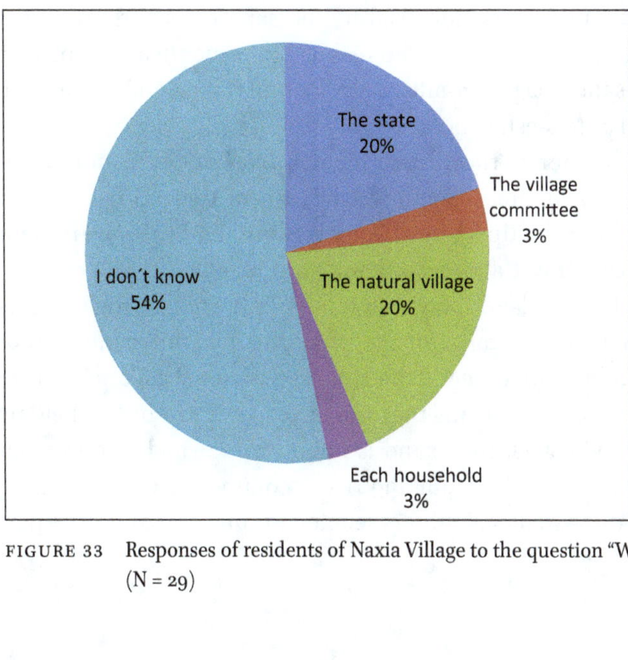

FIGURE 33 Responses of residents of Naxia Village to the question "Who owns river sand?" (N = 29)

who owns the river sand, even though the sand was being sold publicly by the natural village. This reflects the ambiguous ownership of the natural resource, as discussed above. Meanwhile, the majority of the respondents found the protection of sand important.

When asked whether they had heard about any disputes over sand mining, about two-thirds of the respondents said yes. Only one respondent said no, while the others remarkably did not answer the question. It was clear that disputes were taking place during the ongoing river-sand mining. To the question, "Has your village had disputes about sand mining?" ten respondents said "Yes," five respondents answered "No," while "I don't know" was chosen by thirteen respondents. The frequency of the reply "I don't know" reflects, on the one hand, that disputes over sand are a highly sensitive topic that some residents in the research villages do not want to discuss; on the other hand, some who answered "I don't know" were emigrants from Naxia Village who had returned for a visit and were genuinely unaware of the disputes. Furthermore, local people believe that this kind of questionnaire is useless for protecting their interests with regard to river sand and for stopping the illegal river-sand mining. Some villagers told me directly: "It is useless to do this kind of questionnaire." They did not believe that a student could stop illegal sand mining, which has existed for decades. This is the reason many simply replied, "I don't know."

There was constant discussion in the village over who benefited from sand mining, and I therefore designed a question about this topic. Most respondents thought the sand dealers (followed by the sand extractors) were the (main) beneficiaries (see Figure 34). The rest of the respondents pointed to other actors, such as the village committee, the natural village, and the individual villagers, as well as real-estate bosses.

As outlined in chapter 5, the institutional framework for sand governance causes disputes at the village level instead of delivering clear-cut solutions. The legal ownership of sand is ambiguous, and the government has not yet taken consistent action to define and regulate it. The cases of sand-selling in Nazhong Village and Naxia Village indicate that the villagers prefer to believe that they obtain not the use rights but the ownership of the resource, due to the reluctance of the local and national governments to clarify who owns the sand. It seems that from the viewpoint of local elders and the majority of villagers, river sand is not a state-owned resource but a "common-pool resource" (Ostrom 1990) or "common-access resource" (Gordon 1954). In contrast, illegal miners seemed to consider sand to be part

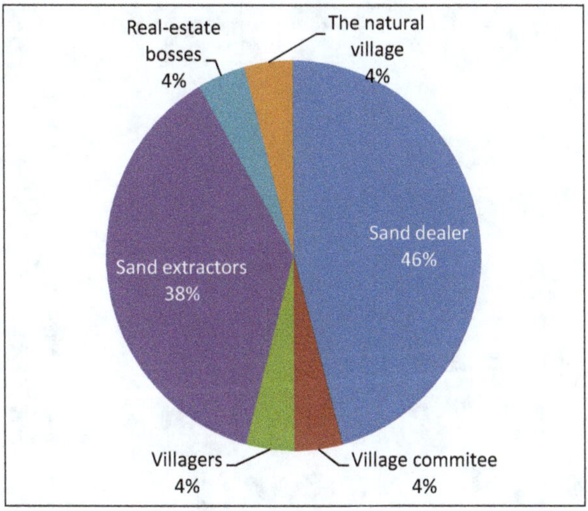

FIGURE 34 Responses of residents of Naxia Village to the question "Who are the beneficiaries of sand mining?" (N = 29)

of an "open-access (no property) regime" (Ostrom 2001), as they claimed, "Sand belongs to nobody." The diverse perspectives on the ownership of river sand, enabled by the ambiguity of the law, have caused disputes in rural areas.

5 An Illegal Sand Agreement and Its Implications

This section describes an agreement that was signed by cadres, party members, and village representatives with a local unlicensed sand miner in Naqiu Administrative Village in Huangwutun Town in 2015. By presenting a longstanding struggle that was caused by an illegal sand agreement, this section further explores villagers' perspectives on ownership rights and rights to the income of river-sand mining.

Figure 35 shows the agreement, which was based on the view that the river sand between Xiadong and Xin Jiankou was owned by the village production teams, who comprise Party A in the agreement. On January 16, 2015, 18 cadres of these production teams and the village representatives signed the agreement with Party B, Mr. Zhang.

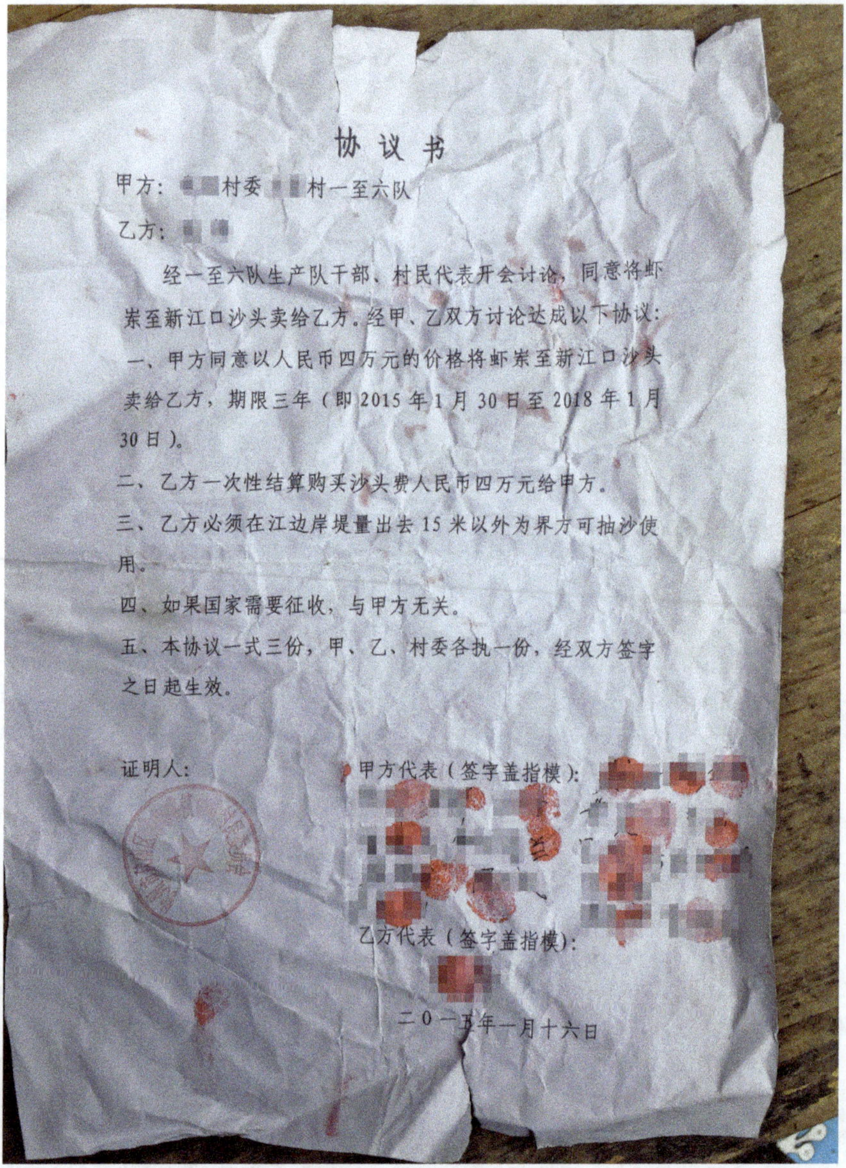

协议书

甲方：村委村一至六队

乙方：

经一至六队生产队干部、村民代表开会讨论，同意将虾宗至新江口沙头卖给乙方。经甲、乙双方讨论达成以下协议：

一、甲方同意以人民币四万元的价格将虾宗至新江口沙头卖给乙方，期限三年（即 2015 年 1 月 30 日至 2018 年 1 月 30 日）。

二、乙方一次性结算购买沙头费人民币四万元给甲方。

三、乙方必须在江边岸堤量出去 15 米以外为界方可抽沙使用。

四、如果国家需要征收，与甲方无关。

五、本协议一式三份，甲、乙、村委各执一份，经双方签字之日起生效。

证明人：

甲方代表（签字盖指模）：

乙方代表（签字盖指模）：

二〇一五年一月十六日

FIGURE 35 The illegal sand agreement
SOURCE: FIELD DATA, 2017

Agreement

Party A: The Six (Production) Teams of Nachun Village, The Committee of Naqiu Administrative Village

Party B: Mr. Zhang

After discussion by cadres and village representatives from production teams 1 to 6, it is agreed that the *shatou* [砂头, the local Zhuang term signifying the location of the river sand or sand beach, or the place where river sand is stored] between Xiadong and Xin Jiangkou is sold to Party B. After the discussion between Party A and Party B, the following agreement was reached.

1. Party A agrees to sell the river sand at a price of RMB 40,000 for a period of three years (from January 30, 2015, to January 30, 2018).
2. Party B pays RMB 40,000 (*shatou* fee) at once to Party A.
3. Party B is not allowed to mine closer than 15 meters from the riverside embankment.
4. If the state needs to levy, it has nothing to do with Party A.
5. This agreement is made in triplicate. Party A, Party B, and the village committee each hold one copy. Effective from the date of signature by both parties.

My informants, including Zeng Pei, the previous captain of one of the production teams, and an elder of Nachun Village, explained that it was Mr. Nong, the party secretary, who organized the meeting with Mr. Zhang, a local villager and a former cadre of the village committee. Mr. Nong conferred with the 18 cadres, the director, and other cadres of the village committee, as well as six villagers who were captains of production teams and also members of the CCP branch of Naqiu Administrative Village. Together they drew up an agreement granting sand-extraction rights to Mr. Zhang for a period of three years (January 30, 2015, to January 30, 2018) in exchange for 40,000 yuan. Mr. Nong received the money and paid each of the 18 cadres 50 yuan for signing the agreement. In order to obtain the signatures of the 18 cadres, Mr. Nong had told them, "The income from selling sand will be used for renovating the public agricultural canals." According to Zeng Pei, the other members of the production teams did not receive anything because the profit would be spent on collective goals, as Mr. Nong had argued.

This agreement did not sit well with the other villagers. It had been signed without informing the majority of the villagers, for the cadres knew very well that the village would not agree to sell the river sand, as I discuss below. Thus, when the villagers found out about the signed agreement, they organized protests to prevent the investor from operating in their river.

Such disputes led to a conflict in which some of Mr. Zhang's workers were injured.

One villager who opposed the agreement was Zeng Pei. He had been farming a specific type of marine worm—*he chong* (禾虫, *Tylorrhynchus heterochaetus*), a regional dish that is also used in Chinese traditional medicine—for years, and he said that the price for *he chong*—80 yuan for 500 grams—was good. Zeng Pei explained that his father, who had been taking care of 20 head of cattle on their field beside the river, discovered the illegal sand mining on January 22, 2015, and that he called Zeng Pei and other villagers to the riverbank on the same day to try to stop the dredging. Mr. Zhang and his workers did not stop, however, telling Zeng Pei and the villagers that they had the right to mine since they had signed a contract with the village committee. The arguments on both sides were clear but diametrically opposed. Violence erupted in the evening, and Zeng Pei called the police in to stop the ongoing sand mining and the violence. Ironically, the police did not stop the sand mining but arrested one of Zeng Pei's uncles, who was accused of injuring the miners. Since then, Zeng Pei and his fellow villagers have drawn up many petitions applying for help from the government. These documents well illustrate the perspectives of the villagers on the ownership, rights, and governance of river sand. I therefore translate one of these petitions here—namely, a request to the Qinzhou Prefectural People's Government made in March 2015 entitled "Requesting the Qinzhou Prefectural People's Government to Stop Mr. Zhang from Illegally Mining River Sand."

1) The *shatou* is the collective property of the production teams.
2) The agreement is illegal. The 18 team cadres and villagers did not convene a village team meeting [i.e., a voting session] before signing the agreement with Mr. Zhang, and the village committee did not convene a party branch meeting to discuss the decision. No written notice was given to the villagers. All of the villagers (more than a thousand people) were unaware of the agreement. Thus, the vast majority of the villagers did not agree to the signing of the agreement. Therefore, this agreement was not signed in accordance with legal procedures.
3) The money is too little. The value of the river sand between Xiadong and Xin Jiangkou is not less than 200,000 yuan. However, the signing of the agreement confirmed a sale of the river sand to Mr. Zhang at a price of 40,000 yuan, and the price was low, resulting in a loss of collective property of the production teams.
4) River sand is a national resource. It was illegal for the 18 cadres and villagers and Mr. Zhang to sign the agreement to sell river sand without the approval of relevant government departments.

5) The river-sand mining area borders a flood-control levee more than one thousand meters long between Xiadong and Xin Jiangkou. The levee plays an important role in flood control and disaster prevention for more than 1,000 *mu* of basic farmland and it also guarantees the safety of the villagers' lives and property. However, there is no provision in the agreement for protecting the levees and basic farmland when mining river sand, resulting in damage to the levees and basic farmland to varying degrees.
6) Article 3 of the agreement stipulates: "Party B is not allowed to mine closer than 15 meters from the riverside embankment." In fact, Mr. Zhang's pumping of sand has caused about 13 meters of collapse between the flood-control embankment and the river, which has seriously jeopardized the safety of the flood levees. Consequently, Mr. Zhang has violated the provisions of Article 3 of the agreement.
7) Article 4 of the agreement stipulates: "If the state needs to confiscate *shatou*, it has nothing to do with Party A [the production teams]." The river sand from Xiadong and Xin Jiangkou is the collective property of the production teams. If the state confiscates the river sand, the control and the right to income [still] belong to the production teams. How is it [i.e., confiscation] irrelevant to the production teams? As a result, all obligations under the terms of the agreement are illegal, and they must be terminated.

The villagers made several recommendations:

> In order to maintain the credibility of the law, protect the collective interests of the production team, ensure the safety of the villagers' production, livelihood, life, and property, and protect the ecological balance, we recommend that the Qinzhou Prefectural People's Government terminate the illegal agreement. We also recommend that the Qinzhou Prefectural People's Government stop Mr. Zhang's illegal sand-mining activity between Xiadong and Xin Jiangkou, and we suggest that the Qinzhou Prefectural People's Government instruct Mr. Zhang to reinforce the damaged flood levee and restore the basic farmland to its original state.

The villagers also planned to take further action themselves:

> We will organize all of the villagers [around one thousand] to petition according to the law in accordance with procedure, allow the media to cover [the event], and use legal weapons to protect our rights if necessary, until the problem is solved.

The villagers strongly assert that the production teams have collective ownership rights to the river sand, but their argument is not entirely consistent. On the one hand, the villagers pointed out that the sandy beach where the river sand is located is a state resource, in order to force the government to take action to stop illegal sand mining. On the other hand, they claimed that even if the state confiscates the river sand, the ownership rights belong to the production teams. They also point out the consequences of sand mining: the loss of land, the collapse of riverbanks, and the severely damaged levee, which is one of the most important flood control facilities in agricultural areas. This response points out the absurdity of Mr. Nong's claim that income from selling sand would be used for improving the canals. The man who had previously been the captain of one of the production teams explained: "When the levee is gone, the whole area will be flooded by sea water. All the canals will be useless" (Nachun Village, March 28, 2017). An elder emphasized: "No villager is allowed to damage the riverside embankment" (Nachun Village, March 28, 2017).

Despite the clarity of the overall argument, the outcome of this and other petitions disappointed the organizers and the majority of the villagers. The prefectural government sent a township government official to ask the party secretary to return the money to Mr. Zhang, and to indicate that Mr. Zhang should stop mining. This happened only because the villagers called the hotline of the mayor of the prefecture several times, in addition to submitting the petition. By that point, however, Mr. Zhang had been mining for an extended period of time (about a month) and had made a great profit. Where did the money go, and what were the consequences for Mr. Zhang? These questions were not addressed by any authority. Eventually, the illegal mining between Xiadong and Xin Jiangkou stopped and the miners moved to other places. No one has been punished, even though the law was violated and the villagers' land and flood-protection levee suffered damage. The villagers did not continue with the petition, nor did they appeal to the court. Instead, they turned their attention to Mr. Nong, the party secretary of Naqiu Administrative Village, and asked for his status to be revoked, a case that will be discussed in detail in the next chapter.

6 Discussion and Conclusion

The case studies from the six Zhuang villages illustrate the villagers' perspectives on sand governance and reflect that both formal and informal institutions have impacted river-sand mining activities and their consequences.

According to Chinese law, neither the village nor any individual household owns river sand because river sand is a state-owned natural resource. However, villagers still claimed that the sand in the river belongs to their villages: sand is considered a common resource, and thus each individual household in a village has a claim to it. Because of these beliefs, for years it was the villagers rather than the government that took action to protect river sand when faced with free-riding and illicit operators in local river sections. The various local reactions reflect that the territory is "the setting for the battles of resistance that impede the implementation of the models of modernity or development" (Toledo, cited in Ulloa 2015, 325).

In rural southwest China, territory refers to the place where one's ancestors settled down, and where family members were born, lived, and worked. It represents the roots of one's family and culture. Moreover, it is a home. Territory is an organizing principle in kinship and religious beliefs. For instance, a territory consists of families living together. A single clan or lineage prefers to live together in one village, even though there are some mixed lineages with diverse family names in some villages in the GZAR's rural areas. People started to identify riparian villages by the surname of a lineage during land reforms in the 1950s and again in the 1980s. Through the lineage-based villages, the villagers highlight the boundaries between diverse villages, using sentences such as: "The land until the bridge is ours; the same applies to the sand." In each village, there are gods of the land, and public land, the river, pools, and mountains constitute the territory. Loughlin (2018, 197) states: "Territory, then, for human beings became more than a piece of earth but was imbued with religious and symbolic importance by the societies that developed on it." A territory is not only a physical phenomenon but is shaped by the memory of the community members and is related to the issue of "our land," the boundary between outsider and insider (see Elden 2013; Ulloa 2015). The members of a village have the right and responsibility to protect their land, houses, and forests, as well as the dikes that protect their properties.

The cases described in this chapter demonstrate that many villagers also consider river sand to be part of a village's territory and it is therefore important to them. River sand relates to the river, water, drought, flooding, dams, fish species, and the enjoyment of children in their territory. Furthermore, the environment of the territory has shaped the people's memories. "When I was a kid, the water in our river was so clear that I could directly see the fish," Zeng Pei told me in March 2017. When Mr. Hu showed me the polluted river water, he said, "When we were herding cattle on the riverside, we drank directly from the water in the river." The villagers in Mei Village said, "We washed vegetables and we washed clothes in the river." River-sand mining in the territory of the rural Zhuang village is influenced by these histories, customs, and local environmental knowledge, as well as concepts of ownership and ideas on environmental justice. For instance, the forbidding of mining closer than 15 meters from the riverside embankment in the three villages in Huangwutun Town stems from local knowledge of flooding and the tidal invasion of sea water. Meanwhile, sharing the profits from the sale of sand at the rate of 2 yuan per participant reflects local ideas on environmental justice and on inclusion of the public in decision-making and benefit sharing.

Interestingly, even the township officers and the administrative village cadres agree on the principle of territory and acknowledge the ownership rights of the natural village. For instance, township officials offered a contract to the village head of Mei Village and asked him to collect the signatures of two-thirds of the households in order to make their sale of sand legal. This implied that other miners were forbidden to mine in the river without the agreement of the villagers. However, the failure of government-based sand control and management has left rural sand mining in chaos. The sand deal between Mei Village and Na Village not only shows an acknowledgment of the rights of villages but also reveals how township officials used their power to benefit from the sand crisis and the government project system. The illegal sand agreement of Nachun Village illustrates the issue of mismanagement by the village committee and the party branch at the administrative level.

The case studies reflect the village's collective power over river-sand mining. The villagers have been able to sell river sand and earn both collective and individual financial income. However, each village has to deal with its own sand crisis and with conflicts both inside and outside its territory. Illegal sand miners, "insiders" and "outsiders" alike, have put formal and informal institutions to the test. This study shows that the commoditization of sand has attracted illegal operators who have begun to extract the resource for sale in the local market. Furthermore, river sand is considered to be part of the collectively owned land and to belong to the production team; because it is not viewed as individual property, individuals have little power to defend their resources.

Moreover, some villages did not want to sell their river sand due to concerns over the ecosystem, flood protection levees, and land protection. As mentioned in chapter 2, the Zhuang people prefer to settle down beside rivers rather than in urban areas. Rivers, water, fish, sand, and land are basic resources for their daily life. In particular, in these agricultural regimes, the link between land and sand is crucial for farming. Water infrastructure, such as flood-protection levees, dams, and irrigation, is key for agriculture in rural areas. Although the villages had a choice between allowing illegal mining and selling their sand, they were unable to stop the mining of their collective resource completely.

In the following chapter, I describe local resistance to illegal sand-mining in the Zhuang villages and explore how the villagers reestablish peace after conflict.

CHAPTER 7

The Sociocultural Consequences of Illegal Sand Mining and Local Resistance

In this chapter, I zoom in on the sociocultural impact of sand-resource marketization. This chapter seeks to answer the questions: What forms does local resistance against illegal river-sand mining take? How do villagers look for justice within the political-cultural structure of their society? And how do the Zhuang villages solve disputes and rebuild peace inside their communities?

Social impacts of the sand-mining industry have been observed around the world, such as in Bangladesh (Khan and Sugie 2015), India (Sekhar and Jayadev 2003; Viju 2009; Safi 2017), Vietnam (Anh 2018), and China (Wu 2008). As discussed in chapter 1, the boom in river sand extraction has received little attention from anthropologists and political ecologists. The political ecology perspective on the consequences of public sand extraction has received less attention in China (Zhu 2018, 2019). When discussing environmental disasters in general, Kirsch (2001, 168–169) argues:

> A distinctive feature of indigenous responses to dispossession and environmental degradation is the claim of "culture loss." ... The notion of loss appears to have two primary registers. It may refer to possession—to the objects or property for which one might claim rights or ownership. Loss in this guise implies value and property relations; it may therefore be possible to gain new understandings of property by examining responses to loss.

Local responses to the loss of river sand reflect how the villagers value their public property, as I discuss in this chapter. Furthermore, it is difficult for the local people to defend their public right to a public natural resource and protect this resource. Kirsch (2001, 168) discusses cultural loss to better understand the notion of "cultural property rights." His argument "is about the disappearance not of entire societies or ways of life but of particular things—knowledge, ideas, and practices of local value." I borrow Kirsch's notion of "cultural loss" to explore how local Zhuang villages experience the loss of river sand, and how they attempt to protect their rights and avoid the further loss of river sand and other resources. In the following section, I focus on two sociocultural domains heavily impacted by sand extraction in the study areas—agricultural

 | DOI:10.1163/9789004505919_008

livelihoods and indigenous belief systems—and on the various ways in which residents resist these changes.

1 Consequences for Agricultural Livelihoods

Livelihoods are a fundamental aspect to study in political-ecological analysis. Livelihood research focuses on how people make their living or achieve well-being by mobilizing their environmental, economic, and social resources (Carr 2015; Hussein 2002; Valdés-Rodríguez and Pérez-Vázquez 2011). One of these resources is river sand. I observed in the Zhuang villages that river sand, which accumulates on the riverbanks in all seasons, supports a wide range of food crops. In particular, due to the scarcity of arable land throughout the region's history, the natural accumulation of river sand is an alternative to artificial beach and wetland reclamation. River sand is one crucial resource related to land, water, and fish for these lineage-based villages, and it has played an important role in the way they make a living. It has, for instance, been used in dams built for agricultural purposes. Moreover, the majority of farmers rely on water from the river for irrigated cultivation and garden cultivation. Results from the household survey among 94 residents of Mei Village (see chapter 2) indicate that the majority of the residents (52 percent) depend on farming for their main livelihood, while the rest of the sampled group mainly engage in off-farm activities, such as wage labor in urban environments.

Farmers usually plant yam seeds in river sand, which is critical to ensure the growth and yield of the crop. The size and physical appearance of Chinese yams (*huaishan*) determine their market value. Mr. Lei, a local farmer from the study area, was adding sand to his yam field beside the main road when I met him in April 2017 (Figure 36). He was carrying a bucket full of sand from a sand heap that was covered with a plastic tent. After I inquired about what he was doing, Lei narrated how the purchase of sand has become the only option for the successful growth of yams:

> I use this sand to grow my yams. These two *mu* (1,333 m^2) of field require about four cubic meters of sand. The market price of yams was 8 yuan per kilogram. Each one of my yams normally weighs between 2.5 and 6 kilograms. In 2016, I earned around 7,000 yuan from selling yams. The price of yams is better than that of rice.
>
> Before 2008, I carried sand from the river onto my plot to grow yams, but this is not possible anymore. In 2009, I began to purchase sand from local vendors at the township sand market. The disappearance of sand

> from the rivers is caused by the massive extraction for the market and the lack of sufficient runoff and floods to carry sand from upstream. We have no option but to buy sand if we want greater yields. Yams that are not grown in sand are of lower quality and are worthless in the market. In the village, at least 30 percent of the 200 households plant yams as a major income-earner. The majority of these households have to buy sand. There are reports that some villagers are stealing sand from fields belonging to others.
>
> Nali Village, April 10, 2017

River sand plays a crucial role in the local cultivation of yams, but it is nowadays only available by importing it from other places, thus increasing the cost of agricultural products. Therefore, it is risky for poor farmers to invest in yams, since yam farming depends on extra-local sand markets. Interestingly, the absence of sand from the rivers has also encouraged the theft of sand from farm plots, which in turn has made growing garden vegetables more difficult. The farmers' difficulties show that the multiple uses of river sand are ever-changing and are shaped by economic and cultural transformations.

FIGURE 36 A farmer storing and transporting sand to grow yams
SOURCE: FIELD DATA, APRIL 2017

Specifically, sand-use patterns have shifted from mostly agricultural purposes to industrial purposes. Carr (2015, 337) linked livelihood research with political ecology and pointed out:

> This framing of livelihoods opens up the analytic consideration of how networks of policy, economy, and power that might result in major livelihoods shifts in particular rural communities are intimately linked to, and productive of, the lives and livelihoods of those we might commonly think of as "powerful."

The overmining of river sand to supply the regional and national construction industry has affected local farms' agricultural processes. An additional factor that has contributed to low yields in affected villages is that of decreasing levels of river water, which complicates the irrigation process. Due to the rapid drop in water levels, irrigation channels can no longer provide a sufficient amount of water. According to Jia et al. (2007, 201), the total volume of sand mined in the lower reaches and the delta of the Dong River in Guangdong Province from 1980 to 2002 was about 332 million cubic meters. This caused the lowering of the average riverbed elevation, an increase in the depth of the water, and the flattening of the longitudinal riverbed gradient (210).

Furthermore, wells and lakes have dried up due to the decreasing water levels of major rivers (see also Mitchell 2006). The lack of irrigation water is further aggravated by damage to water reservoirs caused by mining activities in Huangwutun Town, as observed in a 2019 government report. Moreover, the water is heavily polluted, rendering irrigation untenable for many farmers. Interviews with farmers revealed that some are already turning to rain-fed agriculture and water pumped up from the ground to overcome these issues, but the majority of farmers are poor and cannot afford the use of engine-run water pumps.

In sum, the overmining of river sand has had negative outcomes for communities along the river, such as the collapse of riverbanks, loss of farmlands, damage to dams and bridges, water pollution, river degradation, and loss of fish species and river sand, all affecting the villagers' agricultural livelihoods. These consequences of sand mining have caused tensions on the local level, as I discuss in the following section.

2 Social Conflicts and Local Resistance against Sand Mining

The links between natural resources and conflicts have been extensively studied by social scientists (for instance, Homer-Dixon 1991, 1994, 1999; Robbins

2004, 2012; Billon 2015; Bollig 1990, 1993, 2006; Kioko 2016). An analysis of publications from 2013 showed that "about two-thirds of political ecology studies use the term 'conflict' in their analyses" (Billon 2015, 598). The more specific links between the extractive industries and indigenous peoples have also been studied by scholars (Clark 1996; Gilberthorpe and Hilson 2014; Lertzman and Vredenburg 2005; Whitmore et al. 2012).

Victoria Tauli-Corpuz (in Whitmore et al. 2012, xiii) stated:

> Up to now, at the international arena, when indigenous peoples make their interventions, there are always cases of how extractive industries displaced indigenous peoples from their territories or destroyed their communities. What is more disheartening are the ceaseless reports of continuing violence and grave human rights violations against indigenous peoples in communities where oil, gas or mineral extraction is taking place.

In the case of sand mining in rural Zhuang areas of Qinzhou City, conflicts between the industrial and agricultural uses of river sand and between "winners" and "losers" are the main causes for social tension in the affected areas. The sense of loss of possession is commonly seen in the Zhuang villages in the process of the commercialization of river-sand mining. The loss is visible and has caused numerous disputes and conflicts. This section reflects how sand mining influences local social relations and how local resistance against illegal river-sand mining has taken form. Scott's (1990) term "infrapolitics" is used to illustrate the way that peasants are utilizing strategies of resistance that have gone unnoticed by superordinate governments. As Scott (2012, 113) explains:

> I realized not only that this was a genre of politics but that it was the prevailing genre of day-to-day politics for most of the world's disenfranchised, for all those living in autocratic settings, for the peasantry, and for those living as subordinates in patriarchal families. It was, as it were, a politics that "dare not speak its name," a diagonal politics, a careful and evasive politics that avoided dangerous risks, a politics quite in keeping with the reputation of the peasantry for "cunning."

As far as villagers' resistance in rural China is concerned, O'Brien and Li (2006) believe that the collective rightful resistance of rural citizens toward local governments shows that villagers take rights seriously. In contrast, Pia (2016, 277) demonstrates that "rather than asking to be protected by state laws, people living in the countryside often feel threatened by state rules and legal provisions." My findings illustrate O'Brien and Li's (2006) argument about rural citizens' collective resistance: local villagers, believing that river sand is a common pool resource, viewed the state laws as a weapon and then applied these laws in arguing against illegal river-sand mining and sand mismanagement (see sections 5 and 2.3). Meanwhile, my study also illustrates Pia's argument that rural people may feel threatened by the state: local villagers may distrust the effects of laws on controlling and stopping illegal sand mining (see sections 5.2, 6, 4, and 2.2), or they may feel overlooked when the government fails to respond to villagers' petitions (see sections 4.3, 2.2, 2.3, and 7.2.4), as well as when the county or city government consistently sells off river sections for the mining and privatization of river sand over the course of three decades (see section 5.1).

In the following, I describe three scenarios that exemplify various forms of resistance and one case that illustrates how local government officials and cadres of administrative villages try to deal with villagers' resistance.

2.1 *Resistance through Throwing Stones*

Facing persistent sand extraction, members of Mei, Yang, and Na Villages began to take action to evict illegal sand miners from their localities. On November 3, 2008, a group of people (young and old, women and men) from the three villages assembled close to Mr. Fugeng's dredger and hurled stones at him and his colleagues in the hope that they would flee the area.

The idea of throwing stones at illegal sand harvesters originated in Mei Village, and Mr. Fugeng was the first victim. Following his refusal to vacate the area before November 3, people in this village organized with their neighbors to gather in greater numbers for a serious attack. The elders were able to unite the majority of villagers and the conflict then escalated, which is another indication that people perceive of sand as a common-pool resource that they should defend collectively.

The main point of contention was the fact that a few individuals were making profits from a common resource while the majority of the people did not have the economic power or resources (capital) to venture into the same business. These three villages have long co-existed and shared both water and sand, and they would not sit back and watch a few individuals selfishly benefit from the shared property while most members of the community toiled in the urban environment to earn a "genuine," legitimate wage. Whenever they heard the roaring of a sand dredger, volunteers from these three villages quickly assembled and dashed to the scene to hurl stones at the operators. This form of action was not spontaneous; elders, whom villagers saw as the custodians of their resources, planned and organized the attacks. The largest one took place in 2008.

A woman recalled:

> People from the three villages, including women and children, stood on both sides of the river and began hurling stones at Mr. Fugeng [an insider]. Had he not used a barrel to protect his head, he could have died from the stoning. They all yelled ... "Hit him! Kill him!"
>
> interview in Mei Village, February 9, 2015

During the attack, Mr. Fugeng recognized an old man, Mr. Fulai, who injured him, and he began to orchestrate revenge. He assembled his younger brothers to engage in an attack against Mr. Fulai, who represented the village elders and had acted against illegal sand miners. As the head of Mei Village recalled, Mr. Fugeng succeeded in organizing a larger team—a gang of about 20 men from Xintang Town, most of whom were believed to be drug addicts. Tensions in the village mounted. Someone I informally interviewed in the grocery store narrated the incident:

> Mr. Fugeng's brothers and those affiliated with them armed themselves with knives and sticks to besiege Mr. Fulai's house. Villagers heard them warning anyone who dared to help Mr. Fulai. We were all afraid and did not attempt to intervene.
>
> interview on December 3, 2014

Mr. Fulai himself recalled:

> As for planning, we organized people to stone the illegal sand miners, who were Mr. Fugeng and his partners. The riverbanks were dotted with opponents of illegal sand mining, mainly from Na Village. Soon, I saw that Mr. Fugeng was being stoned by many people, including our villagers. I joined the stone-hurling team while some of our villagers already began to block access to the river using long bamboo sticks.
>
> When I noticed that they wanted to charge at me, I shouted for help and ran toward my cousin's house, where I hid for several days until there was relative calm. Later on, I was informed that Mr. Fugeng and his team had planned to kill all males above 12 years of age in Mei Village in order to prevent any similar situations in the future.
>
> interview in MR. FULAI's home in Mei Village, August 28, 2015

Despite the resistance and violence, the villagers did not solve the problem of illegal sand harvesting, as the illegal sand miners also reacted violently. Moreover, despite their success in uniting some villagers, the elders have lost significant power and influence—hence the illegal sand miners' lack of respect toward or fear of Mr. Fulai.[1]

According to village traditions and in the minds of the local people, all have equal rights to the use of sand as confirmed in the law (de jure rights). Yet the de facto rights to river sand are clearly unequal, because those possessing money, technology, and the ability to intimidate do in fact have much greater access to sand than others. The result of the illegal sand-mining operation was a fracture in the community that caused disharmony and conflict within the affected villages. Sand extraction and related conflicts have affected a once essentially harmonious and unified people. Their long history of socioeconomic cooperation has been increasingly compromised.

Following the failure to stop the illegal sand miners, the elders turned to formal institutions to seek help regarding the governance of sand as a communal resource.

2.2 *Resistance through Appealing to Formal Government Channels*

In early 2017, fearing that the illegal miners would continue to benefit from mining sand while the village would lose out on the benefits of the common resource, members of Naxia Village decided to start selling their sand. However, commercial sand mining by Naxia Village did not proceed without resistance from nearby villages. Indeed, members of other villages under the Naqiu Administrative Village began to raise the alarm, warning that the people from Naxia Village had no right to sell the sand around the Dragon Vein.

Mr. Nong, the party secretary of Naqiu Administrative Village, explained in April 2017 during an interview in his office that some villagers from Nachun had submitted a complaint letter to the government of Naqiu Administrative Village some two weeks prior to the interview. That was three days after Naxia Village had started the village-based sand sale. Mr. Nong immediately took action and called an official of Huangwutun Government in order to report Naxia Village's sand-mining situation. This call bore no fruit, and so Mr. Nong wrote about the situation in Naxia Village on April 7, 2017, in his weekly report to the township government, which, he explained, was a common practice

1 Indeed, the powers of the elders have weakened due to the building of the communist state, for the state has opposed clan authority (elders) in rural areas since the Cultural Revolution in the 1960s. Consequently, the elders have less power regarding the monitoring of disputes, including issues of illegal sand mining.

to communicate the latest information on the rural areas. Two days later, Mr. Zeng, a member of the sixth production team of Nachun Village, filed another complaint about Naxia Village obtaining several thousand yuan from selling river sand. He threatened that if the village committee of Naqiu could not stop this illegal behavior, he too would organize a group of people to sell river sand. Mr. Nong reported to the township government on April 14, 2017, that he had told Naxia Village to stop selling river sand but the organizers had not listened to him. Mr. Nong reminded the government not to delay action any longer, otherwise a disaster would occur due to the other angry villagers in Naqiu Administrative Village. For this reason, he had consistently reported the illegal sand-mining situation to the township government and was now asking for clear instructions. However, he did not obtain a response.

The village elders realized that the government would not take any action with regard to the sand-selling activities. They then started their own sand-selling operation in order to not lose out.

> If you don't sell the river sand, the sand miners might steal it anyway, or the nearby village will take our sand with their sand-selling activities. We had no alternative but to sell the river sand ourselves. The party secretary himself conducted sand-selling activities many times previously. We do not listen to him anymore because this is our sand.
>
> interview with MR. HU and his wife in Naxia Village, April 13, 2017

The lack of response to the village committee's weekly written and oral reports reveals that the government did not dealt with the river-sand mining in a timely manner through any formal institution. Because the township failed to effectively govern rural public resources, people resorted to informal governance. The rural residents, in particular the elders, noted the lack of governmental oversight when it comes to sand mining. Thus, they actively participated in the monitoring of illegal sand mining and took action to reduce the loss of this common good based on their customs and experience.

Similar disputes were occurring nearby among the production teams in Nadong Administrative Village. Interestingly, Nazhong Village has never been reported to the township government for selling sand. What the cadres of Nadong Administrative Village were intending to report to the Huangwutun Township government was another type of conflict: there were disputes between three production teams in one village that required the township government's intervention. The cadres of Nadong Administrative Village were themselves involved in the dispute because Mr. Zhou, the director of the committee of Nadong Administrative Village, was a member of one of the feuding production teams. Mr. Zhou gave the following account of the conflict:

> Currently, there are three production teams struggling with sand mining. The situation is that three of the four production teams—the second (200 members), the third (200 members), and the fourth production team (250 members)—have clarified their boundaries, sold several hundred meters of the river to the sand miners, and obtained several thousand yuan from selling sand. The second and the fourth had sold out their river sand already. The disputes are mainly between the first production team (300 members) and the third. The third production team is going to sell another section of the river to the sand miners. However, the third production team has denied the ownership rights of the first production team by saying that the first production team [to which Mr. Zhou belonged] is located far away from the river and owns no land near the river. In contrast, the first production team argues that three production teams [the first, second, and third] share the same temple of land gods, which is located beside the river. The members of these three teams are brothers and sisters. Furthermore, the first production team previously obtained ownership of land from the Production Brigade during the land reform in 1954, but it gave the management rights over the land to the second team due to the distance between the land and their place of residence. Therefore, the first production team definitely has the same right to benefit as the others. Significantly, the elders of the first team have organized meetings of their team several times; 60 percent of the 70 households have signed a report arguing that the first production team should be recognized as a stakeholder in the river sand. Actually, the first production team only asks to share some benefit from the income, and does not demand that the profits be equally distributed between the two teams. In other words, they argue that the third production team cannot take all the money.
>
> Nonetheless, there is no agreement among these production teams, while the tension is increasing. The cadres of the village committee, including the party secretary and myself, have met to discuss the situation. We think that the disputes have to be reported to the township government to avoid further conflict. As far as I am concerned, I am a member of the first production team and believe that we have the right to share the profits from selling the river sand. Anyway, we have a headache about the inevitable conflict if the village committee and the township government do not deal with the disputes well.
>
> interview on April 12, 2017

Mr. Zhou pointed out that the sand belongs to the production team; the village committee has not obtained any money from sand mining. Mr. Zhou faced controversy since he was not only the director of the village committee but also a member of the first production team.

In short, the cadres on the level of the administrative village attempt to solve disputes with regard to sand mining by applying to formal government channels, but this is commonly unsuccessful and thus not an effective way to deal with resistance against illegal sand mining.

2.3 *Resistance through Petitioning*

As discussed in chapter 6, trust between the cadres and the villagers diminished during the sand-mining crisis and the implementation of projects. The illegal sand agreement and the ensuing conflict in Naqiu Administrative Village in January 2015 serves as an example. In the aftermath, the villagers gathered together to figure out who exactly had made the decision to sell the river sand around the village's flood-control embankment. They concluded that it was the party secretary, Mr. Nong, who had "colluded" with the sand miners and organized an illegal meeting to sell the river sand (see section 5). The villagers discovered that during the meeting, the 18 cadres and village representatives had been forced by Mr. Nong to sign the agreement. Other villagers added that if anyone disagreed with Mr. Nong, there would be negative consequences for them or their household. For example, a cadre was forced to resign after he said that the village committee should not have banquets so frequently but instead should use the money to repair the irrigation canals. In short, villagers believed that Mr. Nong had signed a bad deal on behalf of the village and had not served the common interests. He himself had profited, as had his own production team (see section 4), but he had ignored others' interests in violation of Party rule. The villagers decided to write a report to the leaders of the CCP and the government at the prefectural level with the title: "Report on the Request to Investigate and Deal with the Village Party Secretary Regarding the Illegal Sale of the Village's Flood Control Embankment" (February 3, 2015). The villagers asked the government to force Mr. Nong to resign from his job as the party secretary for the following three reasons.

First, the villagers argued that Mr. Nong had violated Article 3 of the Land Administration Law of the PRC, which stipulates that "no units or individuals may misappropriate, buy, or sell land." Mr. Nong had illegally sold sand to a villager, Mr. Zhang. Meanwhile, according to the "Water Law," it is the state that implements the river-sand mining license system. Mr. Zhang and his group had mined sand without a license. Thus, their behavior constituted a criminal offense.

Second, Mr. Nong had violated Party rules, specifically the rule stating that no major decisions could be taken without organizing a party meeting with more than two-thirds of the party members present. Yet, Mr. Nong had decided to sell the river sand without holding any party meeting.

Third, as the party secretary, Mr. Nong was supposed to consider the livelihoods of the villagers as his first priority. "Focus on people's livelihood, improve people's livelihood, and promote the healthy development of a well-off life in our town," local leaders had been ordered by the Huangwutun Town government. Yet, the villagers pointed out that the illegal agreement disregarded possible consequences for their livelihoods, lives, and even deaths. Once the sand from the village flood-control levee was removed, the levee would collapse, more than 3,000 *mu* of paddy fields would be inundated, and thousands of houses would be flooded. The villagers would then suffer from hunger and homelessness.

For these three reasons, the villagers petitioned the government to dismiss its party secretary and choose another person, one who would wholeheartedly seek to improve the welfare of the villagers. In addition, the prefectural party committee and the prefectural government were requested to investigate and deal with the case of the illegal sale of the flood-control embankment, and furthermore to stop the pumping of sand from the flood-control dikes, to bring the culprits to justice, to assuage the public's anger, and to reassure the people.

Crucially, the people have never given up asking for justice. When I interviewed the organizers of the petition, specifically Zeng Pei and his mother and uncles (one of whom was arrested by the police during the conflict), they gave me their notes written during the process of petitioning. "The government department treats us like a ball, kicking us back and forth," Zeng Pei said. The documents show that the villagers first of all called the party secretary, Mr. Nong, but he said that he was busy and could not come to the riverbank to deal with the conflict. The villagers then called the Land Resources Institute at the township level:

> The Land Resources Institute first asked who was mining, and said they would immediately enforce the law. When the official heard that it was Mr. Zhang, he changed his mind and told the villagers that this was the responsibility of the Water Law Enforcement Brigade. The Water Law Enforcement Brigade pointed out that the township government was in charge of illegal river-sand mining and advised the villagers to report to the government. The government official said that it was the Land Resources Institute's responsibility to control illegal sand mining.
>
> from a report dated February 3, 2015

In short, the fragmented authorities failed to respond to the villagers' complaint.

Zeng Pei recounted his experience with approaching the governmental agencies:

> The competent authorities have treated this big event, which is closely related to the villagers' lives and the property of the whole village, just like a ball to kick around. Each authority kicked the ball to another agency, and they used various excuses to refer the case to each other. The villagers had to send representatives to petition the town government. After three petitions to the town government, it eventually sent a deputy mayor and two cadres to the flood-control levees and they took a brief look for several minutes. However, they said nothing. A few days later, the villagers had not received any response from the township government. Meanwhile, the sand-suckers still smashed the farmland and the flood control levees. The villagers had to send representatives to the town government to petition again. Another official, Mr. Huang, who was the official cooperating with our administrative village, told the villagers that he would call Mr. Nong to give us an answer. The villagers believed that the problem would be solved, based on Mr. Huang's promise. However, there was no feedback. Mr. Huang, in charge of Naqiu Administrative Village on behalf of the township government, is not only the first leader of the village, but also the controller of the matter [the sand agreement]. If he had not been partial to Mr. Nong, there would have been no bloody fighting; however, the bloodshed [the scuffle and the throwing of stones] was inevitable. Therefore, Mr. Huang must take full responsibility for this.
>
> interview in Nachun Village on March 29, 2017

The villagers were very angry and greatly disappointed that the petitions did not stop the illegal river-sand mining. "You can make any petition to any government, even to the prefectural level," Zeng Pei said, repeating what the miner Mr. Zhang had told him and other villagers during the fight. He added:

> I spent so much time making various petitions to the different government bureaus at various levels. But it is pointless. There is no justice at all. The officials are wearing the same pants and they are in collusion. These original documents that we wrote and signed have not been sent to any government bureau because we know that the government does not protect our rights, but only their own interests. What Mr. Zhang said was correct. He and his team had bought the officials all the way up to the prefectural level.

During the focus group discussion in Naxia Village described in chapter 2, the four elders similarly pointed out that the conflict over sand mining in the village had not elicited any feedback from the government. One elder, who is also a CCP member, said that the party members meeting in the village had discussed the new national policy on environmental protection and poverty alleviation. But another elder responded,

> "Serving the people" [a government slogan] was only something they said at the party members' meeting. After the meeting, when they walked out of the meeting room, what they had said in the meeting was abandoned. The party members are mainly thinking about their own pockets. They are thinking about money. They have not contributed to building a better community.
>
> focus group in Naxia Village, March 27, 2017

The government's lack of responsiveness has thus resulted in similar cases of illegal sand mining in other villages.

The villagers believe that there would have been no conflict if the village administrative authority and the township government had dealt with the disputes in a timely and appropriate manner. Facing this reluctance of government officials to take responsibility, the riverfront communities have shifted their focus to social media in order to seek justice.

2.4 *Resistance through Social Media*

Social media petitions are frequently used by people to make their views on illegal sand mining known and to get the attention of the government and the larger society.[2] I argue that villagers use social media as a means to make their grievances known to the public in order to force the government to respond and to enforce the law. Social media here refers to unofficial (non-government) websites, such as regional online platforms, Youku, and Weibo, and chat apps such as WeChat and QQ. Before 2012, people could post anonymously on non-governmental media.[3] Nowadays, more and more governments at diverse

2 On the links between public discussion, information dissemination, and social media in China, see Wallis 2011; Yang 2009.

3 Willnat, Lu, and Martin (2015, 1) state: "In December 2012, the Chinese government toughened its restrictions on social media by issuing new rules that required Internet users to provide their real names to online service providers, while assigning Internet companies greater responsibility for policing controversial content."

levels have established their own social media pages and Internet platforms that people can use to access governmental information and communicate their views and opinions, by registering the users' true names. Foreign social media platforms are blocked in China.

The most well-known local social website in Qinzhou City, "360" (the name of the website indicates the intention to comprehensively report on social and cultural life in the city), has been used for reporting illegal sand mining. Indeed, on December 23, 2017, a search for the term "illegal sand mining" on this website resulted in 26,238 entries. The articles contain evidence, such as photos that depict the sand-mining situation and letters of complaint signed with fingerprints, that reveals the severe consequences of illegal sand mining as well as the anger of the affected rural residents. Moreover, the posts on social media show that sand extraction is taking place all over the main rivers and seas, causing severe ecological consequences and sociocultural disputes throughout the region. Social media petitions display the people's concern with the government's ambiguous position on sand mining and the lack of law enforcement.

Other options for raising awareness of illegal sand mining, besides using social media, are posting on the government website, emailing the secretary of the CCP, and calling the mayor's hotline of Qinzhou City. In particular, investigations by the CCP's Central Environmental Protection Supervision Committee, partly brought about by the complaints of villagers, have forced the local government to act against illegal river- and ocean-sand mining since 2016. In addition, the Qinzhou TV station documented several cases showing the negative consequences of river- and ocean-sand mining in various rivers in the city's territory. However, the justice the villagers are looking for was not achieved easily, even after the report by the TV station.

The concepts of loss, justice, performance of cadres, transparency, and democracy were mentioned by local people when they were talking about petitions. As I discussed previously, there are mainly elders, females, and children left in the villages. Many of them have a telephone and some of them have a smartphone, but few of them know how to submit a petition online. In the case of the illegal sand-mining agreement in Naqiu Administrative Village in 2015, which remained in effect even after the villagers petitioned the township government, Zeng Pei and his relatives uploaded a complaint letter (see Figure 37), signed by about sixty villagers, on a social media page with a large following. The letter was meant to start a conversation on such illegal activities, and it was also another way to petition the government to take action. Zeng Pei informed me that he did not have a smartphone, nor had he used

WeChat before. He purchased a smartphone to learn how to upload his online petition. The cost of rural resistance is high for individual villagers. Zeng Pei explained that he spent a lot of time and money to write the document and asked people for help in making sense of the written complaints. He did not know the law regarding river-sand mining until one of his relatives helped him to formulate the petitions. He himself paid the fees for visiting the various departments of the township and prefectural government. During this period, he was treated like a ball by various authorities. Meanwhile, he could not take care of his agricultural field and fish farm. The situation of Zeng Pei is similar to the circumstances of affected communities located in other "remote areas where education levels are low, and political power is limited, and information restricted," as illustrated by Li and Shapiro (2020, 94). In their case study, the local citizens could not articulate their complaints and organize resistance against the implementation of national renewable energy targets.

Notably, the Nachun villagers were split over the issue of the illegal sand-mining agreement. A good number of them supported the contract, even though the majority opposed it. Eventually, the contract was condemned as "sand misgovernance and illegal cooperation" by the administration, and this ended Mr. Zhang's mining. Nonetheless, the opposing villagers still struggle for usufruct rights, the health of the river, and environmental justice. As Xie (2015, 249) states in a discussion of environmental identity among Chinese environmentalists, "A third facet to environmental identity stems from Chinese citizens' increasing awareness of rights and how these relate to the need to protect the environment."

This in-depth study of a conflict in Nachun Village reveals that villagers have shifted their means of resistance from Scott's (1990) "hidden" style to a "more open" style to protect their common resources. The use of social media is increasingly popular. Xie (2015, 252) argues that "the combination of habitual official intransigence and worsening environmental problems means that public opposition to state policies and practices often bypasses state-sanctioned channels in favor of more direct political action." Therefore, social media gives those who are weak and voiceless a platform to share their opinions and grievances.

The government, however, has its own channels for environmental complaints and petitions. In general, officials only take complaints and petitions seriously if they are made via a government complaint system: the mayor's hotline or a message board for local leaders. The government sometimes responds to specific issues brought up on governmental websites.

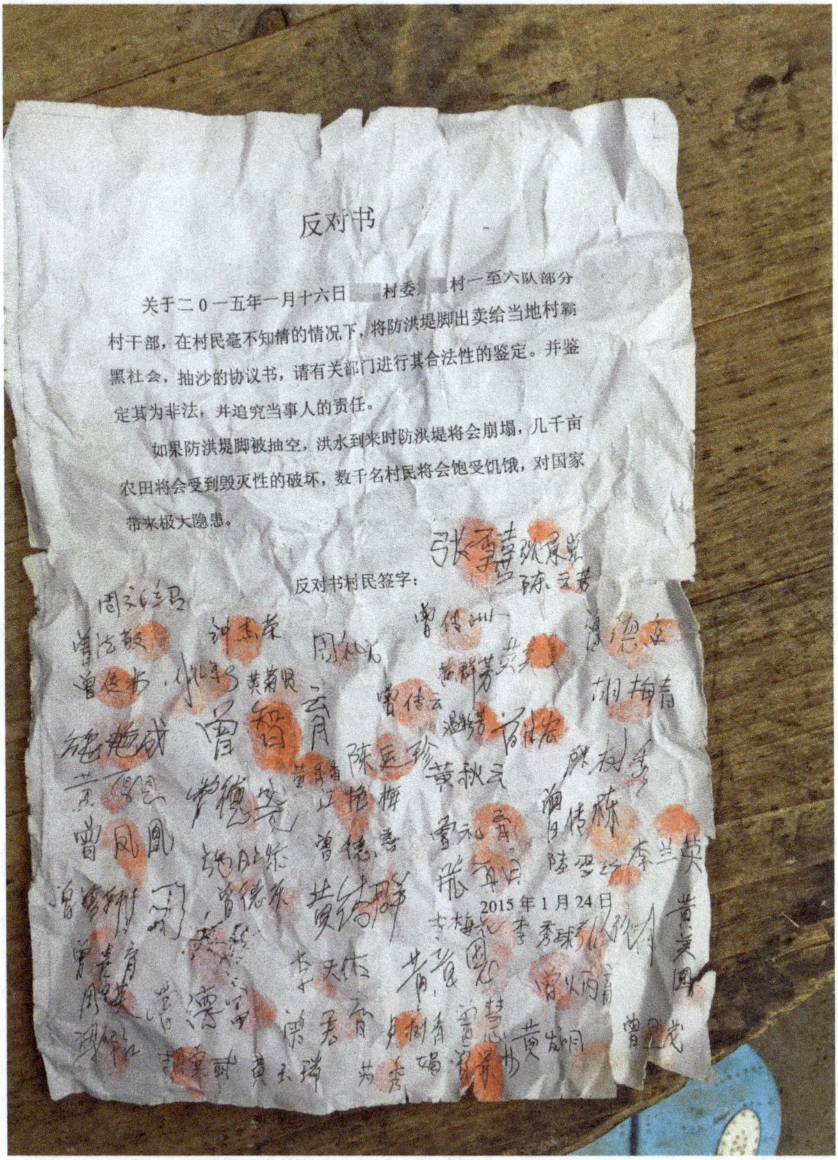

反对书

关于二0一五年一月十六日村委村一至六队部分村干部，在村民毫不知情的情况下，将防洪堤脚出卖给当地村霸黑社会，抽沙的协议书，请有关部门进行其合法性的鉴定。并鉴定其为非法，并追究当事人的责任。

如果防洪堤脚被抽空，洪水到来时防洪堤将会崩塌，几千亩农田将会受到毁灭性的破坏，数千名村民将会饱受饥饿，对国家带来极大隐患。

反对书村民签字：

2015年1月24日

FIGURE 37 Villagers' signatures on the complaint letter. The text of the letter reads: "This is about some village cadres [who signed a contract] which, without informing any major villager, sold the foundations of dams made of sand to local village tyrants on January 16, 2015. We asked the relevant authorities to check the sand-mining contracts and to deem these practices illegal. On top of that, we reminded them that they are legally liable for the major operators' actions. By removing the foundation of the flood dike, several thousand *mu* of farmland could be severely damaged in case of flooding. This would result in thousands of people starving, which in turn represents a serious risk to the county."

SOURCE: FIELD DATA, 2017

3 Sand Exploitation and Its Impact on Indigenous Belief Systems

"People's assets are not merely *means* through which they make a living: they also give *meaning* to the person's world" (Bebbington 1999). More specifically, L'vovich (cited in Sreebha 2008, 1) points out that the long historical interaction between society and rivers is evident in rivers' mystical and religious significance.

There is little research that pays attention to how changes in the environment and ecosystem are impacting villages' river-based belief practices, or to how the villages deal with a tension that exists between the (economic) value of their resources, such as their bridge and river sand, and the (social) value of harmony within their community. My field research reveals that the six riverfront Zhuang settlements in this study, including the way that villagers imagine and describe their boundaries, are deeply embedded in their river environment. The villagers have based their beliefs, food culture, and livelihood on river resources. The customs of festivals, worship activities, and collective fund collections have profoundly informed the common understanding that river-based resources are collective property. This embedded situation forms what Scott (1990) termed a moral economy, which includes customs, property rights, and ideas on belonging.

Lineages play a crucial role in the Zhuang moral economy. Anthropologists have demonstrated the importance of lineages in the formation of sociocultural and sociopolitical spheres (e.g., Freedman 1958; Jing 1996; Siu 1989b; Yang 1996). Freedman (1958) has highlighted how lineage (or clan) organizations have influenced political, social, and economic control in Southwestern China.

I found during fieldwork that these ideas on territory and the Zhuang moral economy generally find consensus among villagers, illegal sand miners, village cadres, and even township government officials. The riverfront natural village is considered to own the river sand located closest to its residents' agricultural land, their temple, their houses, and their territory. This recognition, of course, has shaped the people's sense of common-resource management up until the present. The river-sand crisis raises a significant problem: How do the villages deal with the loss of their common resource and the controversies that ensue inside the village? How can the insider violators be punished, and how can peace and harmony be restored in the village?

Crucially, I found that one key means of governing the common resource and keeping the peace is maintaining collective worship practices and praying together for steady work, wealth, and a happy life. Indeed, religious practices are the major purpose of having a collective fund, apart from constructing water conservation works and other infrastructure. The villagers I interviewed clearly knew the usage patterns of the common fund of their village: "The village rents out the mountain to a private company and obtains 70,000 yuan each year. Therefore, we can hire a better Lingtou team in the festival of Tiao Lingtou" (a sand miner in Nadong Administrative Village, September 2018). As discussed in the introductory chapter, the village-based Lingtou festival is a crucial platform to unite people and reflects a cultural idea of collective religious life. Even the director of the village committee of Nadong Administrative Village described the sand conflicts in his own village by alluding to religious practice:

> We worship the same god of land and practice Zuo She [i.e., worshipping the land gods] together. But the second and the third production team of the village have not distributed the income from selling sand to the first production team. … If our first production team can also share the income of the common sand, then we have some funds for Zuo She.
>
> in the office of Nadong Administrative Village on April 2, 2017

For the villagers, the customary festivals are performed to maintain village solidarity and internal peace among the villagers. Collective identity is formed during these activities through the contributions that villagers make. The donation of money and material as well as voluntary labor has played a big role in many aspects of rural residents' lives. Labor reciprocity is also the major practice in the funeral system in Mei Village and Na Village. Furthermore, each household has the responsibility to work for the Zuo She for free when it is their turn. Some rich villagers donate extra money for collective worship and activities for the Tiao Lingtou festival, in addition to their obligatory contributions. Indeed, the collective fund is crucial to maintaining the belief system.

Interestingly, the illegal sand miners from the villages would also contribute money for the public fund collection: the name of the miner Mr. Fugeng was written on the fund donation list in Mei Village. Because he donated more money and asked for forgiveness from the others, the families of Mr. Fugeng and his brothers have been forgiven by the villagers. Although the villagers forgive the illegal sand miners, they believe that supernatural powers respond to the overexploitation of sand. During my field research, the residents from Mei Village associated the extractive exploitation of sand with an increasing number of unnatural deaths, as well as sickness and related misfortunes. According to them, overexploitation of sand not only ruins sites of sacrifice but also angers the gods. As a result, the villagers consider these misfortunes as a form of supernatural punishment. Three cases exemplify these beliefs.

3.1 *Case One: The Supernatural Punishment of Violators*

Community members believe that illegal sand operations and an elder who supported them have been targets of supernatural punishment. In 2010, while Mr. Fugeng's new house was still under construction, it caved in, which caused huge losses in terms of both material and labor. Indeed, the victim had invested a lot of money in paying for labor because his fellow villagers had refused to assist, as they would have done for other villagers. So long as Mr. Fugeng was being blamed for the illegal extraction of sand, people had not been willing to assist in the construction of his house.

Mr. Fugeng's neighbor commented on the collapse of the house:

> This strange thing never happened in our village before. How is it possible that the roof of the first floor of a house collapses? It is true that a good dog deserves a good bone, while every ill man has his ill day.
>
> interview in Mei Village, September 20, 2015

Other villagers shared these sentiments, acknowledging the role of supernatural powers in Mr. Fugeng's misfortunes. Soon afterward, Mr. Fugeng's father died under mysterious circumstances. The father had supported his sons' illegal mining and it seemed that he was punished for that reason. His death, therefore, was conceived of as a further indication of supernatural punishment. Some villagers connected the death of Mr. Fugeng's father to what the father had done during the clash on December 3, 2018. An informant told me that he was the one who had proposed "to kill all males above 12 years of age in Mei Village." Mr. Fulai, whose responsibility was to organize the funerals in the study area, declined to assist in this case. It was evident that villagers were angry with Mr. Fugeng's family for their involvement in the selfish extraction of sand for commercial purposes.

3.2 *Case Two: The Infestation of Gods' Trees by Insects*

In Zhuang tradition, each village has to have a temple for its land gods. These temples are guarded by a circle of trees, which are considered a form of deity. Coincidentally, these temples are often located on riverbanks from which sand is extracted. In December 2015, some villagers found that insects had feasted on the leaves of some of the trees that surround the temple in Mei Village. This was previously unheard of in the village. Elders interpreted the phenomenon as a form of communication from the gods to show that they were angry. Soon, rumors sprang up that associated this incident and other events, such as the loss of elders discussed in section 3.3, with some of the villagers who had become notorious for commercialized sand harvesting.

Commercial sand mining began to invade the local holy spaces, and conflicts emerged in the research villages. Such conflicts stopped the villagers from performing the land god worship at the temple in Mei Village in 2008. It was evident that the disharmony and disunity that illegal sand harvesting had occasioned, and the related violent conflicts, had created deep mistrust between villagers and exacerbated previously manageable divisions in day-to-day relationships.

Apart from the religious significance of temples and spaces of worship, these places are essential for social interactions. They are considered neutral ground where people share meals and discuss matters of concern relating to the community. It had been a tradition inside most of the villages that heads of household would attend the collective worship of the land god. Labor migration has impacted this collective worship, so that it has become more practical to require a representative of each household, not necessarily the head, to participate in this worship system. As Figure 38 shows, women, young men, and children also participated in the worship in 2014.

The replacement of heads of households by other members is not seen as a reason for the gods to be angry. But the overmining of river sand has physically transformed the river and the land around the river, damaging the Fengshui system of the temple. A religious master explained: "The water pattern changed after the sand was gone. Now, the river water flows directly toward the temple, which is considered to create a bad environment for the temple. The riverbanks beside the temple collapsed. If we cannot stop the illegal sand mining, more land will collapse" (Mei Village, March 30, 2015).

3.3 *Case Three: The Death of Some Elders*

During my fieldwork, the deaths of several villagers were associated with the angering of supernatural powers through the villagers' seeming reluctance to protect holy spaces damaged by sand mining. The villagers were very afraid of punishment from the gods. The damaged templates had to be repaired and the gods had to be worshipped. I attended the funerals of three of the eight victims, during which the presiding priest linked the deaths to the destruction of the temples, which resulted from a relentless pursuit of selfish gains in commercial sand harvesting. Zhu and Qin (2016) pointed out that the traditional funeral has been an informal institution that maintains social integration and is a means of conflict resolution, because the funeral requires the labor of other villagers. Any individual who violates the community norms and customs risks losing out on labor reciprocity at their future funeral.

Figure 39 illustrates funeral traditions and a ritual expressing gratitude for labor reciprocity. The *Shigong* (师公, similar to a priest in Western religion)

FIGURE 38 Sharing an extensive meal under the Gods' trees (*chi tudi*, 吃土地) beside the river
SOURCE: FIELD DATA, 2014

master team was invited to organize the ritual for the funeral depicted in the photo on the left. In addition, other households sent at least one person to help the host to execute the ritual, a common way for the people to deal with death. The photo on the right, taken in 2014, illustrates the ritual performed by the host at the end of the funeral: to demonstrate that the family of the deceased appreciates the helpers, the host offers them money that is supposed to bring good fortune. The ritual reveals the core norms of the rural Zhuang community. Each individual is part of a household, and all households constitute the community. It is not only the lineage group that helps during a funeral or wedding; most of the households in the village community will participate in the ceremony.

These social institutions and the belief system have a strong influence on individuals in Zhuang village society. Together, they answer the question of how villagers value their resources versus how they value members of their community, and how they can make peace again after the sand crisis. The existing social norms and values, including communal harmony and peacemaking, are performed during funerals, weddings, and collective worship, thus

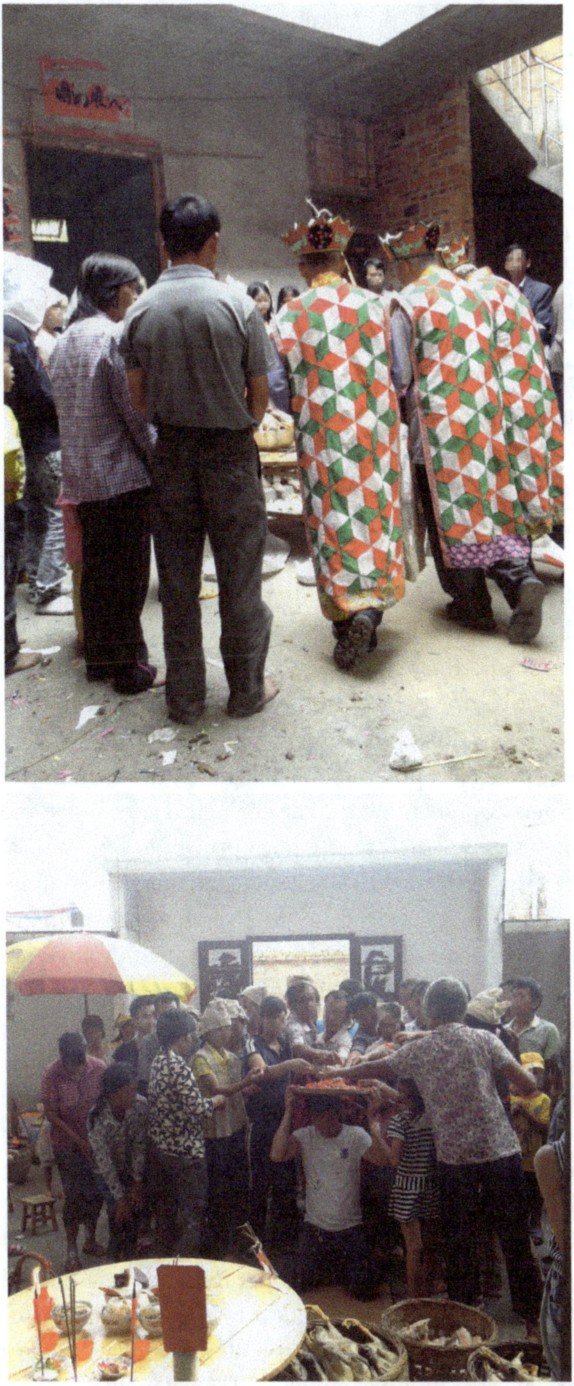

FIGURE 39 Traditional funeral practices (*a*) and a ritual concerning labor reciprocity (*b*)
SOURCE: FIELD DATA, 2015 AND 2014

reinforcing social relationships and the community order. The villagers did not directly take action against a few wrongdoers in their community because the value of village harmony is crucial for the society. Instead, they expected the gods to punish the culprits.

4 Discussion and Conclusion

This chapter has explored the sociocultural consequences of illegal sand mining in the rural Zhuang communities. It shows that conflicts and resistance occur in many riverfront communities in various villages in different ways. It is clear that the villagers believe that river sand belongs to the village or production team. This viewpoint on the ownership of public sand impacts the forms of resistance that villagers practice when they experience the loss of their common property directly. The loss of sand is related to the concept of culture loss. As Kirsch (2001, 167) notes, "The theme of loss has echoes throughout the indigenous world, often in association with damages to and/or displacement from their land." This book demonstrates the process of losing land and associated cultural resources, and the reaction to this loss among the Zhuang people, who depend upon the resources of various ecosystem services for their livelihoods, and as avenues for spiritual nourishment and for group formation and belonging. River-sand mining has put the rural villages on the front line of a natural resource crisis where conflicts emerge. The ecological knowledge and local public management system of the Zhuang have proven to be a crucial factor in the boom of river sand extraction. For instance, the rule that a "miner is not allowed to mine closer than 15 meters from the riverside embankment" reflects that local livelihoods depend deeply on the river system. Apart from this, memories of the river, access to and use rights over river resources, and flooding are all part of the cultural system and, in particular, the belief system. The villagers have also utilized customary patterns in resisting the loss of their sand, such as collecting fingerprints to indicate their support of a petition. This chapter discussed how villagers from three villages took action, even to the point of throwing stones at "thieves." The various collective reactions reflect how the people value river sand as a natural resource.

Despite the tensions related to sand mining, the villages have tried very hard to maintain harmony in their community by consistently practicing funeral labor reciprocity and collective worship. In reaction to the misbehavior of powerful leaders and negative environmental consequences, villagers have filed petitions through formal government channels and social media.

The government's "kicking a ball" approach in dealing with the petitions in Nachun Village has profoundly disappointed the villagers and made the formal sand-management system less effective. The government's refusal of responsibility has resulted in more disputes and violence within rural communities and harmed community–government relations. Consequently, the villagers have to choose a practical strategy, such as selling the river sand to illegal sand miners, to reduce the losses they would suffer if the river sand were depleted without compensation. Thus, the majority of the Mei, Nazhong, and Naxia villagers agreed to sell the river sand in return for a collective income. In Naxia Village, for instance, elders decided to allow the harvesting of sand in areas that are further away from the main infrastructure, such as levees, dams, bridges, and roads. This decision would create some economic returns from the sale of sand while minimizing the extent of infrastructural damage. After the implementation of the river-sand mining licensing system, however, many of the river sections were auctioned by the prefectural or county governments as a way to offer sand on the open market without sharing the profits with the villages, and to govern sand without the participation of the villagers. These complexities arose because of the licensing system.

To sum up, this research shows that many villagers have engaged collectively and individually in protests and resistance against illegal sand mining and have taken action to seek environmental justice, but the local party leaders and responsible government departments have not responded in a positive way, or else have completely ignored them. This chapter has described the subordinate groups' daily resistance as they struggle for resource use rights, the well-being of the river, and environmental justice, using what Scott (1985) termed "weapons of the weak." River-sand management has had specific sociocultural consequences in the six Zhuang villages studied here. Stone-throwing, petitions, and complaints against the party secretary reveal the complexity of the sand-mining crisis at the village level in terms of livelihood, ownership, management, benefits, damage, and environmental impact. Indeed, the agricultural livelihoods of the riverfront villages have been harmed by the loss of river sand, the collapse of land, the pollution of river water, and the lowering of water levels. Furthermore, the belief system based on collective contributions and communal ritual activities has been eroded, since the principles of solidarity and brotherhood were violated.

CHAPTER 8

Conclusion

This book has adopted a political ecology perspective to explore the causes and consequences of illegal river-sand mining in rural Zhuang villages along the Maoling River in southwest China over the last two decades. It has broadened the scope of political ecology studies by focusing on a resource that has recently come into high demand, namely river sand.

Due to the general dearth of publications on river-sand mining and its impacts on Chinese rivers, the complexity and dynamics of the river-sand mining industry in modern China are understudied. The small body of literature on this subject fails to reflect on the political dimensions of human and environment interaction, particularly with respect to China's party-nation power regime. Moreover, studies on the Zhuang minority's culture and local ecological knowledge in the rural areas treated here are also scarce. This scanty literature cannot effectively contribute to current scholarly debates on resource management dilemmas, even though these are prominent in the study area.

This book elaborates on how the governance of sand is both similar to and different from the governance of other resources by discussing its natural characteristics and its uses. The book sheds new light on the sand resource crisis and sand mining management by addressing power relations, institutions, knowledge, narratives, conflicts, resistance, and peacebuilding in the context of the Zhuang communities in southwest rural China. It provides an ethnographic understanding of the interaction of human and river resources, which is a crucial lens for exploring the issues of socioeconomic development and the "looming tragedy of the sand commons" (Torres, Brandt et al. 2017). By taking the Maoling River as an ethnographic subject, this book argues that the river-sand rush and its resulting conflicts reflect the more general present-day ecological crisis and natural resource overexploitation in China, which are driven by the economic-oriented development model. The tension between people over river sand reflects the visible and severe sociocultural and ecological consequences of sand mining.

Prior to the recent sand rush, the uses and values of sand were for a long time ambiguous. Sand has long been considered an abundant natural resource stored in thousands of rivers and lakes. Few people can distinguish the many kinds of sand and its uses. Li (2014, 589) characterizes land by noting: "Land is a strange object. Although it is often treated as a thing and sometimes as a commodity, it is not like a mat: you cannot roll it up and take it away." Unlike

 | DOI:10.1163/9789004505919_009

land, river sand can be taken away by anyone at any time. However, for a long period of time, no one wanted it because it was not useful or valuable. Indeed, the uses of river sand have not been as important as those of land, which has been considered "a source of food, a place to work, an alienable commodity or an object of taxation" (589), or as a resource for amassing capital and power. River sand is not as directly related to people's livelihood as land is. Notably, river sand is largely considered to be either part of the land, or part of the river or the water. In riparian communities, people commonly farm on reclaimed riverbanks, and they normally rely on sand for subsistence—but it is sand in the form of land. This is the case in southwest China, where land is limited and the human population is large, and where sand has been used for farming *he chong* or constructing vegetable gardens. It was not until the last four decades that river sand began to be considered as a basic construction material with a crucial economic value for socioeconomic development. Therefore, sand's economic function has been underlined by an economically oriented government and by the market, while its environmental and sociocultural functions have been undermined.

In the mining era, the environmental and sociocultural consequences of illegal river-sand mining are not fully understood by the government or the sand miners, nor indeed by some villagers. Since there is insufficient environmental evaluation by any authority or third-party environmental agency, the sand miners claim there are no environmental consequences from midstream sand mining. However, the findings of the present study reveal that there are multiple environmental effects—for example, the ecological degradation of the river system, the decline in water level, and the loss of fish species—attributable to river-sand overexploitation in rivers. Significantly, the ecological consequences—hazards, drought, vulnerability, and risk in the riparian communities—still need further exploration.

Recently, environmental research and natural resource management in China has focused mostly on climate change and water pollution (Gilley 2012; Ahlers and Shen 2018; Yamineva and Liu 2019). Although China is the largest producer and consumer of sand in the world (USGS 2013a, 2013b, 2013c; MIIT et al. 2019), little attention has been paid to the commodification of river sand, the ongoing sand rush, and the crisis of river sand. Elsewhere too, river sand is increasingly extracted to support the construction boom, as noted by some scholars and organizations (Torres, Liu, et al. 2017; UNEP 2014; Sreebha 2008; Kondolf 1997; Kondolf et al. 2002; Padmalal et al. 2008). However, these scholars mainly pay attention to the ecological consequences. The sociocultural consequences and diverse actors involved are not given enough attention. Also, the current research on river-sand mining mostly focuses on the weakness of sand-mining institutions and management and pays little attention to ecological issues or the relation between institutions and ecology in China (Wu 2008; Hou 2011). Political ecology is a key tool to understand links between diverse actors by exploring power dynamics and the political, social, and ecological aspects of natural resource management at diverse scales. But political ecology has rarely been used in the context of China.

By applying a political ecology approach to the management of sand, one of the natural resources in highest demand and one that has been marketized to support socioeconomic development, this book contributes to a better understanding of the current sand scarcity and the negative sociocultural and ecological consequences in China and beyond. The book further explores the dynamics of ownership and power regarding access to and control of river sand, providing new insights into modern China's human-river environment relationships, with specific reference to six purposively sampled Chinese Zhuang villages in southwest China. It also contributes an ethnographical study of resource conflict and responds to academic debates on China's natural resource protection and environmental conservation.

The findings show that China is trapped in the dilemma of the common/private–conservation–commodification triad (Prudham 2015, 437) with respect to river-sand mining. I explore a shift in China's river-sand management approach from complete disorder to (an attempt at) complete prohibition. In addition, the findings show that the government as a crucial actor is far from a unified agent. The diverse government departments and governments at all levels have different degrees of authority, and different powers and interests in mining sand. As a result, it is too simplistic to say one government policy for sand mining in China is best, or to only look at either central government or local government action. This kind of dichotomy does not help us to understand the dynamics and complexities of sand's commercialization in the local sociocultural situations. In the mining era, the severe ecological crises of the river system and the overexploitation of river sand have forced governments at various levels to establish new policies or laws. This new kind of formal ecological governance in China is labeled "eco-authoritarianism", or state-led, authoritarian environmentalism, by Li and Shapiro (2020). They further assert: "While many aspects of China's environmental leadership are praiseworthy, other aspects are deeply troubling and should cause us to rethink the wisdom of imagining that a green autocrat can save the planet" (158).

By taking over the management of common resources, the state creates de facto open access because it lacks the financial means to enforce laws and is not able to monitor or exclude immigrant users, as observed by Haller (2010, 413). I have noted the existing contestation between individuals and institutions over the use and management of river sand caused by the vague and contradictory laws. The study finds that the many actors who are involved in the sand-mining process have diverse perceptions about the ownership of river sand, and therefore use various discourses and hold different beliefs about proper strategies for managing sand. The case studies reveal that there are three major actors: the state, villages, and legal and illegal individual miners.

The government holds that river sand is a national resource and implements laws and policies governing it. The state's river-sand management system follows the logic of the interactions between formal institutions on various scales within China's top-down authoritarian system, which seems powerful and effective. But the contradictory laws and the fragmentation of authority between and within different levels of government have led to a diversity of discourses and varying perspectives, increasing rather than decreasing the number of stakeholders and conflicting interests when it comes to sand management. The overlapping laws reflect that river sand is not perceived to be a resource on its own but has been considered to be part of various other resources.

Hardin (1968) states that there are only two solutions to deal with the "tragedy of the commons" in general, namely state ownership or privatization of common resources. China has adopted a combined approach. In practice, the central government with its system of sand-rights auctions aims to maintain the state's control over river sand and to accomplish the privatization and marketization of sand under the sand-mining license system in China. The fact that individuals and companies can obtain sand-harvesting licenses from the government could be thought of as reflecting the privatization approach. Problematically, sand-mining companies are found to have little or no concern for environmental conservation. The scramble for profits outpaces any efforts that are aimed at protecting natural resources. Due to the lack of effective monitoring of licensed mining, the government cannot protect the state-owned resource. Furthermore, the auction system has failed to attract quality miners because the government had difficulties controlling unlicensed sand mining in the rivers.

Since the publication of Garrett Hardin's work in 1968, international scholars have debated what kind of institution (democratic regime or authoritarian management), what kind of philosophy (Western rationality or indigenous ecological knowledge), and what scale of governance (state level or community level) can better preserve the environment and reduce carbon emissions to save the planet. Criticism of the purpose and effect of China's authoritarian environmentalism is part of this discussion. It responds to Li and Shapiro's arguments about the problems of state-led authoritarian environmentalism in China. They argue that "a more accurate term than 'authoritarian environmentalism' is 'environmental authoritarianism,'" since they found that authoritarianism is the end and environmentalism only the means (Li and Shapiro 2020, 24).

Apart from the government, the local communities, which in this case are composed of a collection of Zhuang villages, also claim rights to sand. Their claim that "it is our sand" stems from the Zhuang villages' settlement pattern and communal resource customs and traditions. For communities in the rural GZAR, the management of river sand then becomes a type of common-pool resource management. Ostrom (1990) points out that common-pool resource management only succeeds under certain circumstances. The issue in this case is that the illegal miners (both insiders and outsiders) cannot be excluded. Instead of empowering the local community to fully govern the river sand, the state has heavily intervened in the community's common resource governance.

Furthermore, "illegal" sand miners also claim rights to land. Their illegality relates to their lack of sand-mining licenses and the fact that they often ply their trade in the dead of the night. These "sand poachers" do not obtain permission from villagers. For such actors, sand is an open-access resource. It is this multiplicity of actors and institutions that makes sand extraction and management a complex, rushed, and dangerous affair.

The findings show that both formal (sand-license system and government-managed river-sand mining) and informal institutions (lineage customs and traditional communal resource governance) have engaged in river-sand mining mainly through the "sale" of rights. It is worth pointing out that there are many struggles between formal and informal institutions in terms of ownership, authority, and interests in resource management in modern China (see also Siu 2016; Ho 2015). Both formal and informal channels seem to be weak avenues to prevent illegal sand mining, even though an analysis of the use of Internet platforms has shown citizens' awareness of the consequences of sand mining. Formal institutions have also been transformed in order to deal with the sand crisis. Political reform is taking place and new laws are being enacted by the government at various levels in order to achieve the national goal of ecological conservation (Xi 2017). Indeed, public natural resource management and the construction of ecological civilization in China is mainly an administrative-based and top-down reform that is embedded in the authoritarian system (Li and Shapiro 2020; Liu et al. 2019b; Han and Lai 2012). This focus on the so-called "rise of authoritarian environmentalism" (Li and Shapiro 2020, 1) reflects that many global scholars understand China mainly from a political institutional perspective. However, there are multiple perspectives from which to explore China's ecological civilization construction and environmental management, including cultural, social, and economic institutional perspectives, as well as the future direction of China. A multi-perspective analysis contributes to sketching a bigger picture of present-day China and its global implications. Furthermore, China has only taken about three decades to achieve its current economic accomplishments. It is necessary to mention that the rise of authoritarian environmentalism and the construction of ecological civilization have only recently begun. Scholars should analyze the influence and outcome of "environmentalism with Chinese characteristics" from a future perspective and both within China and beyond. For example, more anthropological ethnography on the environmental crisis and climate change in China should be done for this purpose.

The fragmentation of authorities between and within different levels of government has not led to good governance over resources but has created the opposite situation, in which there is collective-leadership management in theory but no agency takes concrete responsibility in practice. Furthermore, the strong market demand for natural resources due to the construction boom encourages the market actors to collude with autocratic agencies. Inevitably, corruption and mismanagement occur among government officials due to local and departmental protectionism, and there is a misuse of governmental power in sand-mining management. This situation has resulted in a huge waste of public sand, which has severe sociocultural and ecological consequences. Even though the central government's ecological supervision system has been strengthened, this supervision system is temporary and cannot solve the management problems in the long run (see also Li and Shapiro 2020; Liu et al. 2019). The government reports of Huangwutun Town show the failure of the local river chief to stop illegal mining after the implementation of the River Chief System. This system adds an environmental conservation target to the government at all levels by making river protection a political responsibility. In 2018, sand mining was banned in an increasing number of rivers due to severe environmental degradation and social unrest. Nonetheless, illegal sand mining still exists and has become a headache for the government after the implementation of the River Chief System.

It is difficult to fully understand the future implications of the rapid economic development and increasing urbanization rate in China without discussing in detail the exploitation of natural resources and the increase in rural–urban migration. This book has explored how illegal river-sand mining plays a crucial role in regional construction development driven by the government. Illegal river-sand mining is an important part of the sand commodity chain because it has to support rapid economic growth, industrialization, and urbanization. Thus, the resource of sand has made great contributions to the establishment of China's ever-increasing economy. However, the level of development is highly uneven within provinces, cities, and even villages, which is an important topic for analysis when considering the migrant workers and the translocal sand commodity chain. Both sand and migrant labor are cheap and are transferred from rural to urban areas. Rapid urbanization has in some areas consumed most of the river sand. In some cases, the building of new houses solely for immediate demolition to gain compensation has led to a huge waste of river sand and labor, as I discussed in chapter 4.

Furthermore, the River Chief System is an administrative-based reform meant to overcome the fragmentation in water management (Liu et al. 2019), not one based in law. Laws are largely ignored in China's current political reform. The central government highlights the effectiveness of administrative-based reform in terms of river conservation and sand management. Even though some regional laws have been enacted, the enforcement of laws is weak, and laws are not fundamentally applied in resource management in China. Interestingly, laws have been used by diverse actors in various circumstances to justify their actions. Laws have therefore figured in many "blame games" in everyday debates and in the media. The case studies in Huangwutun Town have shown how the villagers labeled the sand agreement as illegal and applied laws in order to force the government officials to stop illegal sand mining. The exclusion of other marginalized stakeholders has weakened the implementation of national resource policy. For instance, villages and villagers who once depended on a rural common-pool resource for making a living are excluded from the government's decision-making, monitoring, law enforcement, and benefit-sharing processes with regard to sand management. The exclusion of stakeholders and an administrative rather than law-based approach makes river-sand management ineffective and unsustainable.

In the case of rural Zhuang villages, dilemmas in the management of public sand and the increasing number of conflicts in relation to the marketization economy show that villagers have recognized the difficulties in managing common-pool resources. "The legitimacy of excludability is critical" in local autonomous institutions for managing common-pool resources (Choe and Yun 2017, 113). The Zhuang communities cannot effectively exclude the insiders and outsiders who mine sand to benefit themselves instead of the community. When the villagers want to protect the collective right to the common resource or, if they fail, try to reduce their loss, they use varying discourses. As described in this book, they claim their ownership by highlighting the concept of territory and the relation between sand, the river, and land. They convene mass meetings, make use of the elders' traditional authority and moral discourses, and even throw stones at illegal miners. In order to achieve an agreement inside the group and to build a united leadership to fight against the ones who "steal" their sand, or to sell the sand in a collective way, they need to establish an informal institution to govern their common resources, even though there is not a united pattern of village-based sand management and the solutions differ per village. The common absence of state power in riverfront villages contributes to the boom of illegal sand mining, to the over-harvesting of sand, to the destruction of the riverine environment, and to the development of related conflicts, while mismanagement by local officials creates distrust between the township government and local communities.

In order to reduce the loss of the common resource of sand and to ask for justice, villagers engage in various forms of local resistance, such as exposing cases of illegal mining on social media, making complaints to the authorities, and calling the hotline of the mayor while referring to state laws and Party rule. However, demands for action are treated like a hot potato by the responsible authorities. The justice that local people are looking for is not achieved easily, even after their struggles have been reported by the media. The reluctance of government officials to take responsibility made governmental sand management highly ineffective. The various social conflicts and forms of local resistance reflect the extent of the sand crisis in rural Zhuang areas, where illegal sand mining has existed for three decades without effective government control.

Based on insights from the field of political ecology, I have highlighted some of the power dynamics between different stakeholders and institutions at various scales, as well as the disputes arising between them. I have identified some of the "winners" and "losers" in sand resource marketization. The "winners" have obtained big profits from river-sand mining through their access to power and money, but most members of the riverfront communities have lost out and only have access to a limited share of the profits, if at all. They now have to live beside rivers drained of sand and suffer the multiple consequences of sand overmining. In the long term, the Zhuang villages along the Maoling River are severely affected by the rampant river-sand extraction in terms of villagers' territory, customs, agricultural production, and daily activities.

This book finds that the villagers had little power to act against illegal sand mining. There were no local township or county authorities that had power to enforce policies. Illegal sand miners clearly figured out the power dynamics in sand management. They bribed government officials in order to access sand without paying the (higher) auction fee, and they signed contracts with villages allowing them to mine in the river within the villages' territory. It seems that these stakeholders all benefited from the sand industry, but my research has shown that the Zhuang villagers did not want sand mining in their territory. In order not to lose out completely when mining occurred, however, they made collective efforts to be recognized as stakeholders by miners, village cadres, and local officials. After achieving this recognition (mostly in an informal way), the villagers were sharing a 2- to 8-yuan profit per cubic meter from public sand mining, as I discussed in chapter 6. Their share of the profit was very small. The illegal sand miners, wholesalers, truck owners, and government officials took most of the profit when river sand became commodified.

The case studies reveal that the local resource ideology and the indigenous ecological knowledge system play a role in the varied forms of informal sand management in the Zhuang villages. The villagers have considered the consequences of sand mining for the river water, dams, irrigation canals, riverbanks, riparian land, water levels, and the overall ecological system of the river. They frequently cited historical droughts and floods and their consequences, and they are afraid of the potential damage to their livelihoods and the environment due to the problematic exploitation of sand.

Even though the villagers' struggles and resistance are informed by the Zhuang minority's culture and ecological knowledge, this knowledge is not recognized in formal river conservation. For instance, the "15-meter prohibition line" in the sample villages in Huangwutun Town that is meant to protect the levees is not recognized by the miners or enforced by the government. I observed that the villagers cannot protect the levees, and I therefore disagree with some scholars who have argued that indigenous knowledge may contribute to solving the crisis of natural resource management and environmental degradation in rural areas inhabited by ethnic minorities (Duara 2015; Senanayake 2006; Aswani, Lemahieu, and Sauer 2018; Anderson 1996). These works have not explored how, in the ensuing processes of negotiation, compromise, dispute, and conflict, the territory has been changed physically and discursively. I thus conclude that the riverfront Zhuang villagers and their ecological knowledge have proven to be less powerful than government policy, economic forces, and developmental trends. Local ecological knowledge is of limited relevance in a context of great power disparities.

The book also provides an example of an "anthropology of home." It is based on 13 months of ethnographic fieldwork in my home area, in the course of which I was able to engage with relevant stake holders, including government officials and local cadres, villagers and their representatives, and (illegal) sand miners, wholesalers, and ship and truck workers. The high demand for river sand has made the increasing scarcity of the shared resource all too clear, but the market-oriented extraction continues unabated, with serious implications at the local level.

I felt frustrated when the villagers explained how hard they had tried to procure justice by approaching government authorities and media, to no avail. Some informants questioned whether my research was useful for them. The villagers and I shared feelings of frustration about the ongoing illegal sand mining and were disappointed by the failure of township and county officials to govern sand mining. Even some township and city officials shared this disappointment and asked me for advice. Illegal sand mining seems unstoppable in this region.

A sound sand-management solution does not seem to be forthcoming. The phenomenon of unlicensed rural sand mining could be described as the result of a strange combination of state socialist and neoliberal approaches to commodity resources. This special type of state socialism imposes its top-down power structure on sand-mining management. However, it does not provide an effective approach for governing river sand in riverfront Zhuang areas because of the overwhelming marketization of sand extraction. The state stimulates the marketization of river sand, while the local government uses auctions to sell mining rights to private companies.

Fieldwork showed that the power structure is dynamic. Apart from the party-state's formal structure, informal institutions and other powerful actors have significantly influenced sand-mining activities. Varying levels of power and unequal resource rights have resulted in the production of various discourses on sand mining: tragic narratives, development narratives, and sustainable narratives. Indeed, beliefs about spiritual punishment have been applied in an attempt to solve conflicts and disputes. The revival of ritual is a way to reconstruct social harmony in rural Zhuang communities.

If we bring these points together, we gain a clearer picture of illegal sand mining in rural Zhuang villages in Southwest China. This study has shown the development of the commodity chain of river sand in the region due to the construction boom in rural and urban areas, which is promoted by governmental and private investments. As a result of the mismanagement of river-sand mining by the government, legal and illegal sand miners have heavily overexploited sand. River sand is not considered as a subject that has intrinsic value but as an object for sale. River resources have not only supported the daily economic and cultural needs of people, but they also have a crucial value for non-human beings, including aquatic life and the riparian ecological system. The governance of river-sand mining reflects the overall resource crisis and degradation of the environment in China. The threat of sand overextraction, the loss of fish species, the ongoing degradation of rivers, the rise of sociocultural struggles and conflicts, and the transformation of state policies together all lead to a rather sobering conclusion. "China must truly 'go green,' " as argued by Li and Shapiro (2020, 205).

This book has identified the riverfront communities as the "losers" in river-sand mining. However, from a long-term perspective, there is no "winner" once all minable sand is gone from rivers, lakes, and the ocean. This raises important questions about the sustainability of sand-mining practices in China. Conservation is in the interest of all, but it is the riparian people who care most about the ecological situation of the river. Therefore, when new policies and laws concerning river sand are being formulated, the Zhuang communities, as well as other ethnic groups who settle down along rivers, should be part of regional or national decision-making.

References

AGWRHT. 2007–2017, 2020. "Huangwutunzhen renminzhengfu gongzuo baogao" 黄屋屯镇人民政府工作报告 [Annual Government Work Report of Huangwutun Town]. Huangwutun Town: The People's Government of Huangwutun Town.

AGWRXT. 2010–2017, 2020. "Xintangzhen renminzhengfu gongzuo baogao" 新棠镇人民政府工作报告 [Annual Government Work Report of Xintang Town]. Xintang Town: The People's Government of XinTang Town.

Ahlers, Anna L., and Yongdong Shen. 2018. "Breathe Easy? Local Nuances of Authoritarian Environmentalism in China's Battle against Air Pollution." *The China Quarterly* 234: 299–319.

Anderson, E. N. 1996. *Ecologies of the Heart: Emotion, Belief, and the Environment.* New York: Oxford University Press.

Anh, Ngan. 2018. "Illegal Sand Mining Digs Away at Vietnam's Rice Basket." *Vietnam Express*. March 19. https://e.vnexpress.net/news/business/illegal-sand-mining-digs-away-at-vietnam-s-rice-basket-3724777.html (accessed April 15, 2018).

Aswani, Shankar, Anne Lemahieu, and Warwick H. H. Sauer. 2018. "Global Trends of Local Ecological Knowledge and Future Implications." *PLoS ONE*. April 5. doi.org/10.1371/journal.pone.0195440 (accessed April 15, 2018).

Atkinson, A. 1991. *Principles of Political Ecology.* London: Belhaven Press.

Axinn, W. G., and L. D. Pearce. 2006. *Mixed Method Data Collection Strategies.* New York: Cambridge University Press.

Bagnold, R. A. 1954. *The Physics of Blown Sand and Desert Dunes.* London: Methuen.

Bai, Zhili, and Juan Liu. 2020. "China's Governance Model and System in Transition." *Journal of Contemporary East Asia Studies* 9: 65–82.

Bailey, S., and R. L. Bryant. 2005. *Third World Political Ecology: An Introduction.* London: Routledge.

Barnes, J. 2014. *Cultivating the Nile: The Everyday Politics of Water in Egypt.* Durham, NC: Duke University Press.

Barnier, Michel. 2003. "Guidelines for Successful Public-Private Partnerships." *Website of the European Commission.* https://ec.europa.eu/regional_policy/sources/docgener/guides/ppp_en.pdf (accessed March 20, 2019).

Bebbington, A. 1999. "Capitals and Capabilities: A Framework for Analyzing Peasant Viability, Rural Livelihoods, and Property." *World Development* 27: 2021–2044.

Beeson, Mark. 2010. "The Coming of Environmental Authoritarianism." *Environmental Politics* 19: 276–294.

Beeson, Mark. 2016. "Environmental Authoritarianism and China." In *The Oxford Handbook of Environmental Political Theory*, edited by Teena Gabrielson, Cheryl Hall, John M. Meyer, and David Schlosberg, 520–532. Oxford: Oxford University Press.

Bel, G. 2011. "Infrastructure and Nation Building: The Regulation and Financing of Network Transportation Infrastructures in Spain (1720–2010)." *Business History* 53: 688–705.

Bernard, H. R. 2006. *Research Methods in Anthropology: Qualitative and Quantitative Approaches*. Lanham, MD: Altamira Press.

Bilik, Naran 纳日碧力戈. 1990. "Minzu yu minzu gainian bianzheng" 民族与民族概念辨正 [Dialectic of Minzu and the Minzu Concept]. *Minzu yanjiu* 民族研究 [Ethno-National Studies] 5: 11–38.

Bilik, Naran 纳日碧力戈. 1995. "Minzu yu minzu gainian zai bianzheng" 民族与民族概念再辨正 [Re-dialectic of Minzu and the Minzu Concept]. *Minzu yanjiu* 民族研究 [Ethno-National Studies] 3: 9–16.

Billon, Philippe Le. 2015. "Environmental Conflict." In *The Routledge Handbook of Political Ecology*, edited by Tom Perreault, Gavin Bridge, and James McCarthy, 598–608. London: Routledge Taylor & Francis Group.

Blaikie, Piers. 1985. *The Political Economy of Soil Erosion in Developing Countries*. London: Longman.

Blaikie, Piers. 1995. "Changing Environments or Changing Views? A Political Ecology for Developing Countries." *Geography* 80: 203–233.

Blaikie, Piers, and Harold Brookfield. 1987. *Land Degradation and Society*. London: Methuen.

Bollig, Michael. 1990. "Ethnic Conflicts in North-West Kenya: Pokot-Turkana Raiding 1969–1984." *Zeitschrift für Ethnologie* 115: 73–90.

Bollig, Michael. 1993. "Intra-and Interethnic Conflict in Northwest Kenya: A Multicausal Analysis of Conflict Behaviours." *Anthropos* 88: 176–184.

Bollig, Michael. 2006. *Risk Management in a Hazardous Environment: A Comparative Study of Two Pastoral Societies*. New York: Springer.

Bridgea, Gavin, Begüm Özkaynakb, and Ethemcan Turhanc. 2018. "Energy Infrastructure and the Fate of the Nation: Introduction to Special Issue." *Energy Research & Social Science* 41: 1–11.

Bryant, Raymond. 1992. "Political Ecology: An Emerging Research Agenda in Third-World Studies." *Political Geography* 11: 12–36.

Bryant, R. L., and S. Bailey. 1997. *Third World Political Ecology*. London: Routledge.

Bryceson, D. F., 1996. Deagrarization and rural employment in sub-Saharan Africa: A sectoral perspective. *World Development*, 24(1), p. 97–111.

Bumpus, A. G., and D. Liverman. 2008. "Accumulation by Decarbonization and the Governance of Carbon Offsets." *Economic Geography* 84: 127–155.

Cai, Kevin G. 1999. "Outward Foreign Direct Investment: A Novel Dimension of China's Integration into the Regional and Global Economy." *The China Quarterly* 160: 856–880.

Callicott, J. B. 1994. *Earth's Insights: A Multicultural Survey of Ecological Ethics from the Mediterranean Basin to the Australian Outback*. Berkeley: University of California Press.

Carr, Edward R. 2015. "Political Ecology and Livelihoods." In *The Routledge Handbook of Political Ecology*, edited by Tom Perreault, Gavin Bridge, and Jamens McCarthy, 332–342. London: Routledge.

Cen, Shu-Tian 岑树田. 2019. *Daguo jishi: zhongguo jichu sheshi de fazhan luoji* 大国基石：中国基础设施的发展逻辑 [The Cornerstone of a Big Country: The Development Logic of China's Infrastructure]. Guangxi: Guangxi Normal University Press.

Chan, Kam Wing. 2014. "Achieving Comprehensive Hukou Reform in China." Paulson Institute. *Paulson Policy Memorandum*. December.

Chatterjee, P., 1983. "More on Modes of Power and the Peasantry." *Subaltern Studies* 2: 311–349.

Chen, Baifeng 陈柏峰. 2011. "Cunzhuang jiufen jiejue: zhuti yu zhiquan — Wancun Getangcun diaocha" 村庄纠纷解决：主体与治权——皖村葛塘村调查 [The Settlement of Village Disputes: The Subject and the Right to Govern: A Survey of the Village of Gudun in Anhui Village]. *Dangdai faxue* 当代法学 [Contemporary Law Review] 5: 22–31.

Chen, Hui-min, and Yun-bao Xu. 2017. "Research on the Mechanism of Ecological Resources Impoverishment." In *2nd International Conference on Sustainable Energy and Environment Protection (ICSEEP 2017)*, 31–35. Changsha, China: DEStech Publications.

Chen, Jie, Juan Jing, Yanyun Man, and Zan Yang. 2013. "Public Housing in Mainland China: History, Ongoing Trends, and Future Perspectives." In *The Future of Public Housing*, edited by Jie Chen, Mark Stephens, and Yanyun Man, 13–35. Berlin: Springer.

Chen, Lei 陈雷. 2004. "Chenlei zai Changjiang hedao caisha guanli biaozhanghui shang de jianghua" 陈雷在长江河道采砂管理表彰会上的讲话 [Speech at the Awards Ceremony for Sand Mining Management in the Yangtze River]. Website of the Ministry of Water Resources of the People's Republic of China. http://www.mwr.gov.cn/slzx/slyw/200412/t20041221_147441.html (accessed December 7, 2017).

Cheng, Si. 2018. "Illegal Berths, Sand-Mining Targeted Along Yangtze River." *China Daily*. September 28. http://www.chinadaily.com.cn/a/201809/28/WS5bad7f52a310eff30327feb0.html (accessed November 6, 2018).

Cheng, Zhiming. 2014. "The New Generation of Migrant Workers in Urban China." In *Urban China in the New Era: Market Reforms, Current State, and the Road Forward*, edited by Zhiming Cheng, Mark Wang, and Junhua Chen, 125–153. Wollongong: Springer.

Chien, Shiuh-Shen, and Dong-Li Hong. 2017. "River Leaders in China: Party-State Hierarchy and Transboundary Governance." *Political Geography* 62: 58–67.

Choe, Hyun, and Sun-Jin Yun. 2017. "Revisiting the Concept of Common Pool Resources: Beyond Ostrom." *Development and Society* 46: 113–129.

Clark, A. 1996. "Economic, Social and Cultural Issues in Mineral Development in Vietnam." *Asia Pacific Management Review* 4: 177–191.

Collins, B. D. 1991. "River Geomorphology and Gravel Mining in the Pilchuck River." Snohomish County, Washington. Report prepared for the Pilchuck River Coalition, 30 pp.

Collins, B. D., and T. Dunne. 1986. "Gravel Transport and Gravel Harvesting in the Humptulips, Wynoochee, and Satsop Rivers." Grays Harbor County, WA. Report prepared for Grays Harbor County Planning and Building Department, Montesano, WA, 70 pp.

Collins, B. D., and T. Dunne. 1987. "Assessing the Effects of Gravel Harvesting on River Morphology and Sediment Transport: A Guide for Planners." Report prepared for the State of Washington Department of Ecology, Olympia, WA, 45 pp.

Corson, C., and K. I. MacDonald. 2012. "Enclosing the Global Commons: The Convention on Biological Diverity and Green Grabbing." *Journal of Peasant Studies* 39: 263–283.

COSC (Census Office of the State Council 国家统计局人口司) and DPESNBS (Department of Population and Employment Statistics 国家统计局人口和就业统计司). 2012. *Zhongguo 2010 Nian renkou pucha ziliao* 中国2010年人口普查资料 [Tabulation of the 2010 Population Census of the People's Republic of China]. Beijing: China Statistics Press.

Court of Qinbei District. 2015. "Shi mou danwei xinghui yishen xingshi panjueshu" 石某单位行贿案一审刑事判决书 [Criminal Judgment on the First Charge in the Bribery Case of Mr. Shi]. Website of China judicial document. May 27. https://susong.tianyancha.com/addf87781cb811e6b554008cfae40dc0 (accessed August 5, 2018).

CPRC04. 2004. *Zhonghua renmin gongheguo xianfa* 中华人民共和国宪法 [The Constitution of the People's Republic of China]. Beijing: The National People's Congress.

CPRC54. 1954. *Zhonghua renmin gongheguo xianfa* 中华人民共和国宪法 [The Constitution of the People's Republic of China]. Beijing: The National People's Congress.

CPRC82. 1982. *Zhonghua renmin gongheguo xianfa* 中华人民共和国宪法 [The Constitution of the People's Republic of China]. Beijing: The National People's Congress.

Csordas, Thomas J. 1988. "Embodiment as a Paradigm for Anthropology." *Ethos* 18: 5–47.

Csordas, Thomas J. 1993. "Somatic Modes of Attention." *Cultural Anthropology* 8: 135–156.

CSSA (China Sand-Stone Association 中国砂石协会), CR (CementRen 水泥人网), and CC (ChinaConcretes 中国混凝土网). 2018. "Jinji, duosheng wusha kemai" 紧急！多

省市无砂可买 [Urgent! Many Provinces Have No Sand to Buy]. http://mini.eastday.com/mobile/180822201800888.html (accessed December 27, 2018).

CSSGLW (Gravel Aggregate 砂石骨料网). 2018a. "Duodi quanmian jinzhi hedao caisha! 'Yidaoqie' neng guanzhu feifa tousha ma?" 多地全面禁止河道采砂！"一刀切"能管住非法偷砂吗？ [Multi-site Comprehensive Prohibition of River Sand Mining! Can "One-Size-Fits-All" Control Illegal Sand Stealing?]. March 27. http://www.cssglw.com/News/news_content/id/927363678441818.html (accessed December 27, 2018).

CSSGLW (Gravel Aggregate 砂石骨料网). 2018b. "Jinkou hesha xin huoyuan: Dongdiwen shuyi dun hesha zhunbei xiang zhongguo chukou" 进口河砂新货源：东帝汶数亿吨河砂准备向中国出口 [New Sources of Imported River Sand: Hundreds of Millions of Tons of River Sand in East Timor are Ready to Export to China]. October 16. http://www.cssglw.com/News/news_content/id/916774900165992.html (accessed December 5, 2018).

Davis, Aeron, and Catherine Walsh. 2015. "The Role of the State in the Financialisation of the UK Economy." *Political Studies* 64: 666–682.

Deng, Yang-Kui 邓艳葵. 2012. "Zhuangzu chuantong shengtai lunli jiazhi tanjiu" 壮族传统生态伦理价值探究 [An Inquiry into the Value of the Traditional Ecological Ethics of the Zhuang Ethnic Group]. *Guangxi minzu daxue xuebao* (*zhexue shehui kexue ban*) 广西民族大学学报（哲学社会科学版） [Journal of Guangxi University for Nationalities (Philosophy and Social Science)] 34(2): 102–104.

Diamond, J. 2011. "China, Lurching Giant." In *Collapse: How Societies Choose to Fail or Succeed*, by J. Diamond, ch. 12. New York: Penguin Books.

Dong, Wen-Feng 董文锋 and Yi-Ze Chen 陈贻泽. 2018. "Dishierjie zhongguo (Nanning) guoji yuanlin bolanhui kaimu" 第十二届中国（南宁）国际园林博览会开幕 [Opening of the 12th China (Nanning) International Garden Expo]. December 6. http://www.gxzf.gov.cn/sytt/20181206-725377.shtml (accessed February 9, 2019).

Duan, Changqun 段昌群 Xueqing Yang 杨雪清 and Wenyi Zhang 张文逸. 2000. "Shengtai huanjing wenti dui xin xin zhongguo zhengzhi shenghuo zhi yingxiang: cong zhengzhi shengtai xue de jiaodu fenxi" 生态环境问题对新中国政治生活之影响:从政治生态学的角度分析 [Effects of Eco-Environmental Issues on Political Affairs in China During the Past 50 Years ——A Perspective of Political Ecology]. *Sixiang zhanxian*思想战线 [Thinking] 04: 44–47.

Duara, Prasenjit. 2015. *The Crisis of Global Modernity: Asian Traditions and a Sustainable Future.* Cambridge: Cambridge University Press.

Eaton, Sarah, and Genia Kostka. 2018. "What Makes for Good and Bad Neighbours? An Emerging Research Agenda in the Study of Chinese Environmental Politics." *Environmental Politics* 27: 782–803.

Economy, Elizabeth C. 2004. *The River Runs Black.* Ithaca, NY: Cornell University Press.

Economy, Elizabeth C. 2007. "The Great Leap Backward? The Costs of China's Environmental Crisis." *Foreign Affairs* 86: 38–59.

Edin, Maria. 2003a. "Remaking the Communist Party-State: The Cadre Responsibility System at the Local Level in China." *China: An International Journal* 1: 1–15.

Edin, Maria. 2003b. "State Capacity and Local Agent Control in China: CCP Cadre Management from a Township Perspective." *The China Quarterly* 173: 35–52.

Elden, S. 2013. *The Birth of Territory.* Oxford: Oxford University Press.

EPLPRC14. 2014. *Zhonghua renmin gongheguo huanjing baohufa* 中华人民共和国环境保护法 [The Environmental Protection Law of the People's Republic of China]. Beijing: Standing Committee of the National People's Congress of the People's Republic of China [全国人民代表大会常务委员会].

EPLPRC89. 1989. *Zhonghua renmin gongheguo huanjing baohufa* 中华人民共和国环境保护法 [The Environmental Protection Law of the People's Republic of China]. Beijing: Standing Committee of the National People's Congress of the People's Republic of China [全国人民代表大会常务委员会].

Erskine, W. D. 1988. "Environmental Impacts of Sand and Gravel Extraction on River Systems." In *The Brisbane River: A Source-book for the Future,* edited by W. D. Erskine, 295–303. Brisbane: The Australian Littoral Society in association with the Queensland Museum.

Expertenkommission. 2015. Expertenkommission für Forschung und Innovation. https://www.leopoldina.org/ueber-uns/kooperationen/expertenkommission-forschung-und-innovation/ (accessed July 1, 2021).

Escobar, Arturo. 1996. "Construction Nature: Elements for a Post-Structuralist Political Ecology." *Futures* 28: 325–343.

Falzon, Mark-Anthony. 2009. *Multi-Sited Ethnography: Theory, Praxis and Locality in Contemporary Research.* London: Routledge.

Faure, David, and Helen F. Siu. 1995. *Down to Earth: The Territorial Bond in South China.* Stanford, CA: Stanford University Press.

Fazekas, Mihály, and Bence Tóth. 2017. *The Effectiveness of the European Union in Safeguarding Competition in Public Procurement Markets.* Budapest, Hungary: Government Transparency Institute.

Fei, Xiaotong. 1939. *Peasant Life in China: A Field Study of Country Life in the Yangtze Valley.* New York: E. P. Dutton.

Fei, Xiaotong 费孝通. 1998. *Xiangtu zhongguo shengyu zhidu* 乡土中国生育制度 [Rural China and Its System of Child-Bearing]. China: Beijing University Press.

Feng, Le-kun 冯乐坤. 2009. "Cunmin zizhi lifa zhi pipan" 村民自治立法之批判 [Critics on Villagers' Autonomy]. *Fazhi yu shehui fazhan* 法制与社会发展 [Law and Social Development] 15(1): 133–141.

Fidjett, S. 2003. *River Mining: Planning Guidelines for the Management of River Mining in Developing Countries with Particular Reference to Jamaica.* Birmingham, UK: Alliance Environment and Planning Ltd.

Forsage, A., and N. E. Carter. 1973. "Effects of Gravel Dredging on the Brazos River." *Proceedings of the Annual Conference of the Southeastern Associations of Fish and Game Commissioners* 27: 695–709.

Forsyth, T. 2008. "Political Ecology and the Epistemology of Social Justice." *Geoforum* 39: 756–764.

Foster, Robert J. 2005. "Tracking Globalization: Commodities and Value in Motion." In *Handbook of Material Culture*, edited by Christopher Tilley, 285–302. London: SAGE.

Freedman, Maurice. 1958. *Lineage Organization in Southeastern China.* London: Athlone Press.

Freedman, Maurice. 2007. *Chinese Lineage and Society: Fukien and Kwangtung.* London School of Economics Monographs on Social Anthropology 33. Oxford: Berg.

Fu, Guang-Hua 付广华. 2014. *Zhuangzu diqu shengtai wenming jianshe yanjiu – jiyu minzu shengtaixue de shijiao* 壮族地区生态文明建设研究——基于民族生态学的视角 [Research on Ecological Civilization Construction in the Zhuang Nationality: Based on the Perspective of Ethnic Ecology]. Guilin: Guangxi Normal University Press.

Galay, V. J. 1983. "Causes of River Bed Degradation." *Water Resources Research* 19: 1057–1090.

Galero, D. M., H. E. Pesci, and P. J. Depetris. 1998. "Effects of Quarry Mining and of Other Environmental Impacts in the Mountainous Chicam-Toctino Drainage Basin (Cordoba, Argentina)." *Environmental Geology* 34: 159–166.

Gauke, David. 2016. *HM Treasury Budget 2016.* London: The Williams Lea Group.

GBSYC (People's Government of Guangxi Zhuang Autonomous Region 广西壮族自治区人民政府). 2011. "Zhuang zu" 壮族 [The Zhuang Minority]. http://www.gxzf.gov.cn/mlgx/gxrw/ftrq/20110402-285607.shtml (accessed December 11, 2018).

GBSYC. 2015a. "Guangxi shiju minzu gaishu" 广西世居民族概述 [Overview of Guangxi Historical Residents]. http://www.gxzf.gov.cn/mlgx/gxrw/ftrq/20150204-437609.shtml (accessed June 18, 2018).

GBSYC. 2015b. "Yu yan" 语言 [Language]. http://www.gxzf.gov.cn/mlgx/gxrw/ftrq/20150205-437672.shtml (accessed March 24, 2018).

GBSYC. 2018. *Guangxi Overview.* http://en.gxzf.gov.cn/2018-08/17/c_263420.htm (accessed September 14, 2018).

Giddens, A. 1984. *The Constitution of Society: Outline of the Theory of Structuration.* Cambridge, UK: Polity Press.

Gilberthorpe, Emma, and Gavin Hilson. 2014. *Natural Resource Extraction and Indigenous Livelihoods: Development Challenges in an Era of Globalization.* Ashgate, UK: Routledge.

Gilley, Bruce. 2012. "Authoritarian Environmentalism and China's Response to Climate Change." *Environmental Politics* 21: 287–307.

GMP. 2005. "Nanning International Convention and Exhibition Center, China." https://www.gmp-architekten.com/projects/403/nanning-international-convention-and-exhibition-center-china (accessed December 23, 2017).

GOGZARG (General Office of the Guangxi Zhuang Autonomous Region Government 广西壮族自治区政府办公厅). 2017. "Guangxi zhuangzu zizhiqu quanmian tuixing hezhangzhi gongzuo fangan" 广西壮族自治区全面推行河长制工作方案 [Fully Implement the River Chief System in Guangxi Zhuang Autonomous Region]. Working plan. Guangxi: GOGZARG.

Golding, Wyatt. 2011. "Incentives for Change: China's Cadre System Applied to Water Quality." *Pacific Rim Law and Policy Journal* 20: 399–428.

Goodale, Mark. 2017. *Anthropology and Law: A Critical Introduction*. New York: New York University Press.

Gordon, H. S. 1954. "The Economic Theory of a Common Property Resource: The Fishery." *Journal of Political Economy* 62: 124–142.

GOSCPRC (General Office of the State Council of the PRC 国务院办公厅). 2007. "Guowuyuan bangongting guanyu zhuanfa nongyebu cunmin yishiyiyi chouzi choulao guanli banfa de tongzhi" 国务院办公厅关于转发农业部村民一事一议筹资筹劳管理办法的通知 [Notice of the Management Method of "One Project, One Discussion" about Raising Funds and Soliciting Labor from Villagers, The General Office of the State Council]. 2007, no. 4. Beijing: General Office of the State Council.

GOSCPRC (General Office of the State Council of the PRC 国务院办公厅). 2016. "Guanyu quanmian tuixing hezhangzhi de yijian" 关于全面推行河长制的意见 [Policy Opinion Regarding Full Implementation of the River Chief System]. *Guowuyuan bangongting tongzhi* 国务院办公厅通知 [Circular of the General Office of the State Council]. Beijing: General Office of the State Council of the People's Republic of China.

Greenberg, J., and T. Park. 1994. "Political Ecology." *Journal of Political Ecology* 1: 1–12.

Guo, Zhenglin 郭正林 and Daming Zhou 周大鸣. 1996. "Nongmingong waichu yu nongmin xiandaixing de huode" 农民工外出与农民现代性的获得 [Emigrant Workers and the Acquisition of Modernity by Peasants]. *Zhongshan daxue xuebao (shehui kexue ban)* 中山大学学报（社会科学版） [Journal of Sun Yat-sen University (Social Science Edition)] 5: 53–58.

Gustafsson, B., and D. Sai. 2009. "Villages Where China's Ethnic Minorities Live." *China Economic Review* 20: 193–207.

Haller, T. 2010. "Between Open Access, Privatisation and Collective Action: A Comparative Analysis of Institutional Change Governing Use of Common-pool Resources in African Floodplains." In *Disputing the Floodplains: Institutional Change and the Politics of Resource Management in African Wetlands*, edited by T. Haller, 413–443. Leiden: Brill.

Hammerschmid, Gerhard, and Kai Wegrich. 2016. "Infrastructure Governance and Government Decision-making." In *The Governance Report 2016*, edited by the Hertie School of Governance, 31–54. Oxford: Oxford University Press.

Han, Haoying. 2012. "National Land Use Management in China: An Analytical Framework." *Journal of Urban Management* 1: 3–38.

Han, Longxi 韩龙喜, Wei Li, 李伟, Yongjun Lu 陆永军, and Shiping Mo 莫诗平. 2005. "Rengong washa dui Dongjiang shuixi shuidongli ji huanjing yinxiang fenxi" 人工挖沙对东江水系水动力及环境影响分析 [Impact of Artificial Sand Excavation on the Hydrodynamics and Water Environment of Dongjiang River Network]. *Hehai daxue xuebao (ziran kexue ban)* 河海大学学报(自然科学版) [Journal of Hohai University: Natural Sciences] 2: 123–126.

Hardin, Garrett. 1968. "The Tragedy of the Commons." *Science* 162: 1243–1248.

Hargrove, E. 1986. *Religion and Environmental Crisis*. Athens: University of Georgia Press.

Harrell, Stevan. 1995. *Cultural Encounters on China's Ethnic Frontiers*. Seattle: University of Washington Press.

Harvey, M. D., and T. W. Smith. 1998. "Gravel Mining Impacts on San Benito River, California." *Proceedings of International Water Resources Engineering Conference, Hydraulics Division* 1998: 304–309.

Hathaway, Michael J. 2010. "The Emergence of Indigeneity: Public Intellectuals and an Indigenous Space in Southwest China." *Cultural Anthropology* 25: 301–333.

Hathaway, Michael J. 2016. "China's Indigenous People? How Global Environmentalism Unintentionally Smuggled the Notion of Indigeneity into China." *Humanities* 5: 54. doi:10.3390/h5030054.

He, Xuefeng 贺雪峰. 2013. *Xin xiangtu zhongguo (xiudingban)* 新乡土中国（修订版）[Rural China in the 21st Century]. 3rd ed. Beijing: Beijing University Press.

Hecht, S. B., and A. Cockburn. 1992. "Realpolitik, Reality and Rhetoric in Rio." *Environment and Planning D: Society and Space* 10: 367–375.

Hershkovitz, Linda. 1993. "Political Ecology and Environmental Management in the Loess Plateau, China." *Human Ecology* 21: 327–353.

Ho, Peter. 2001. "Who Owns China's Land? Policies, Property Rights and Deliberate Institutional Ambiguity." *The China Quarterly* 166: 394–421.

Ho, Peter. 2005. *Institutions in Transition: Land Ownership, Property Rights, and Social Conflict in China*. Oxford: Oxford University Press.

Ho, Peter. 2006. "Trajectories for Greening in China: Theory and Practice." *Development and Change* 166: 3–28.

Ho, Peter. 2015. "Myths of Tenure Security and Titling: Endogenous, Institutional Change of China's Development." *Land Use Policy* 47: 352–364.

Holifield, R. 2015. "Environmental Justice and Political Ecology." In *Handbook of Political Ecology*, edited by Tom Perreault, Gavin Bridge, and James McCarthy, 585–597. London: Routledge.

Homer-Dixon, Thomas. 1991. "On the Threshold: Environmental Changes as Causes of Acute Conflict." *International Security* 16: 76–116.

Homer-Dixon, Thomas. 1994. "Environmental Scarcities and Violent Conflict: Evidence from Cases." *International Security* 19: 5–40.

Homer-Dixon, Thomas. 1999. *Environment, Scarcity, and Violence.* Princeton: Princeton University Press.

Hong, Nai-Feng 洪乃丰. 2002. "Haisha dui gangjin hunningtu de fushi yu duice" 海砂对钢筋混凝土的腐蚀与对策 [Sea Sand Corrosion of Reinforced Concrete and Countermeasures]. *Hunningtu* 混凝土 [Concrete] 8: 12–14.

Hong, Nai-Feng 洪乃丰. 2004. "Luyan fushi yu gangjing zuxiuji" 氯盐腐蚀与钢筋阻锈剂 [Chloride Corrosion and Rebar Inhibitor Admixture]. *Hunningtu* 混凝土 [Concrete] 1: 58–60.

Hong Kong International Airport, 2016. "Hong Kong's new three runway system." August 1. https://www.threerunwaysystem.com/en/information-centre/press-release/01082016/ (accessed November 11, 2018).

Horowitz, Leah S. 2015. "Local Environmental Knowledge." In *The Routledge Handbook of Political Ecology*, edited by Tom Perreault, Gavin Bridge, and James McCarthy, 235–248. New York: Routledge.

Hou, Liangze 侯良泽. 2011. "Hedao caisha guanli gongzuo zhong cunzai de wenti he duice" 河道采砂管理工作中存在的问题和对策 [Existing Problems and Their Countermeasures While Managing River Sand Mining]. *Zhongguo shuili* 中国水利 [China Water Resources] 10: 27–34.

Hou, Liqiang. 2018. "River chiefs keep water bodies clean". *The State Council of the People's Republic of China*. July 18. http://english.www.gov.cn/state_council/ministries/2018/07/18/content_281476226902566.htm (accessed March 12, 2018)

Howell, Anthony J. 2017. "Impacts of Migration and Remittances on Ethnic Income Inequality in Rural China." *World Development* 94: 200–211.

Howell, Anthony, Björn Gustafsson, and Sai Ding. 2018. "Investigating the Patterns and Determinants of Han and Ethnic Minority Household Migration in China." In *Ethnicity and Inequality in China*, edited by Björn A. Gustafsson, Reza Hasmath, and Sai Ding, 259–282. New York: Routledge.

Huang, Bing-Sheng 黄秉生. 2003a. "Zhuangzu wenhua genxi yu zhuangzu wenhua shengtai mei — Zhuangzu wenhua shengtai mei yanjiu zhi yi" 壮族文化根系与壮族文化生态美——壮族文化生态美研究之一 [Cultural Roots and the Ecological Beauty of the Zhuang Nationality Culture — A Study on the Ecological Beauty of the Zhuang Nationality Culture (No. 1)]. *Guangxi minzu xueyuan xuebao (zhexue shehui kexue ban)* 广西民族学院学报(哲学社会科学版) [Journal of Guangxi University for Nationalities (Philosophy and Social Sciences Edition)], 2: 81–86, 103.

Huang, Bing-Sheng 黄秉生. 2003b. "Guangna wenhua genxi yu Zhuangzu de wenhua shengtai mei— Zhuangzu wenhua shengtai mei yanjiu zhi san" 广纳文化根系与

壮族的文化生态美—壮族文化生态美研究之三 [Absorbent Root Systems and the Ecological Beauty of the Zhuang Nationality Culture—A Study on the Ecological Beauty of the Zhuang Nationality Culture (No. 3)]. *Guangxi minzu xueyuan xuebao* (*Zhexue shehui kexue ban*) 广西民族学院学报(哲学社会科学版) [Journal of Guangxi University for Nationalities (Philosophy and Social Sciences Edition)]. 4: 98–101.

Huang, Bing-Sheng 黄秉生. 2003c. "Zhuangzu wenhua shengtai de jingsheng zhi mei—Zhuangzu wenhua shengtai mei yanjiu zhi liu" 壮族文化生态的竞生之美——壮族文化生态美研究之六 [The Beauty of Rivalry in the Zhuang's Cultural Ecology: Series of Studies on the Ecological Beauty of the Zhuang's Culture (No. 6)]. *Guangxi minzu xueyuan xuebao* (*Zhexue shehui kexue ban*) 广西民族学院学报(哲学社会科学版) [Journal of Guangxi University for Nationalities (Philosophy and Social Sciences Edition)]. 5: 92–96.

Huang, Haoming 黄浩铭. 2016. "Guangxi chutai hedao caisha guanli tiaoli daji gelei feifa caisha xingwei" 广西出台河道采砂管理条例打击各类非法采砂行为 [Guangxi Promulgated River Sand Mining Management Regulations to Crack Down on Illegal Sand Mining]. http://www.gov.cn/xinwen/2016-12/01/content_5141449.htm (accessed March 12, 2018).

Huang, Q. D., and J. J. Xu. 2019. "Rethinking Environmental Bureaucracies in River Chiefs System (RCS) in China: A Critical Literature Study." *Sustainability* 11: 1608. https://doi.org/10.3390/su11061608 (accessed January 2, 2020).

Huang, Xian-Fan 黄现璠. 1962. "Nongzhigao qibing fansong shi zhengyi de zhanzheng" 侬智高起兵反宋是正义的战争 [Nong Zhigao's Fight against the Song Dynasty Was a Just War]. April 2. *Guanxi ribao* 广西日报 [Guangxi Daily].

Huang, Zongzhi 黄宗智, Weigang Gong 龚为纲, and Yuan Gao 高原. 2014. "Xiangmuzhi de yunzuo jizhi he xiaoguo shi helihua ma" '项目制'的运作机制和效果是'合理化'吗? [Is the Operation Mechanism and Effect of the "Project System" "Rationalized"?]. *Kaifang shidai* 开放时代 [Open Times] 5: 143–159.

Huangma Assets 皇马资产. 2019. "Qinbeiqu Qinjiang, Maolingjiang bufen heduan hedao caishaquan xiangmu huanjing yingxiang pingjia diyici gongshi" 钦北区钦江、茅岭江部分河段河道采砂权项目 环境影响评价第一次公示 [The First Public Announcement of the Environmental Impact Assessment of the Project of Sand Mining Rights in Some River Sections of Qin River and Maoling River in Qinbei District]. http://huangmazichan.com/news-datail.php?id=291 (accessed January 2, 2020).

Hussein, K. 2002. *Livelihoods Approaches Compared: A Multi-Agency Review of Current Practice.* London: Department for International Development.

Jackson, J. E., and K. B. Warren. 2005. "Indigenous Movements in Latin America, 1992–2004: Controversies, Ironies, New Directions." *Annual Review of Anthropology* 34: 549–573.

Jahiel, A. 1998. "Organization of Environmental Protection in China." *The China Quarterly* 156: 757–787.

Jia, Liang-Wen 贾良文, Zhang-Ren Luo 罗章仁, Qing-ShuYang 杨清书, Su-Ying Ou 欧素英, and Ya-Ping Lei 雷亚平. 2007. "Daliang caisha dui Dongjiang xiayou ji Dongjiang sanjiaozhou hechuang dixing he chaoxi dongli de yingxiang" 大量采沙对东江下游及东江三角洲河床地形和潮汐动力的影响 [Impacts of the Large Amount of Sand Mining on Riverbed Morphology and Tidal Dynamics in the Lower Reaches and Delta of the Dongjiang River]. *Dili xuebao* 地理学报 [Journal of Geographical Sciences] 61(9): 197–211.

Jiang, Wenran. 2009. "Fuelling the Dragon: China's Rise and Its Energy and Resources Extraction in Africa." *The China Quarterly* 199: 585–609.

Jiang, Wenyue 蒋文跃. 2015. "Qinbeiqu shuiliju yuan juzhang Laohongzhi shouhui 177 Wan, zuo gongjiao shangban" 钦北区水利局原局长劳宏智受贿177万 坐公交上班. [The Former Director of the Qinbei District Water Conservancy Bureau, Lao Hongzhi, Took Bribes of 1.77 Million and Took a Bus to Work]. http://www.gxnews.com.cn/staticpages/20150403/newgx551dcbfe-12524621.shtml (accessed March 20, 2018).

Jiao, Yuanmei, Xiuzhen Li, Luohui Liang, Kazuhiko Takeuchi, Toshiya Okuro, Dandan Zhang, and Lifang Sun. 2012. "Indigenous Knowledge and Natural Resource Management in the Cultural Landscape of China's Hani Terraces." *Ecological Research* 27: 247–263.

Jing, J. 1996. *The Temple of Memories: History, Power, and Morality in a Chinese Village.* Stanford, CA: Stanford University Press.

Jordan, Thomas D. 1980. *A Hand Book of Gravity-Flow Water Systems.* Kathmandu, Nepal: United Nation's Children's Fund.

Kanehl, P., and J. Lyons. 1992. *Impacts of Instream Sand and Gravel Mining on Stream Habitat and Fish Communities Including a Survey on the Big Rib River.* Research Report 155. Marathon County, Wisconsin: Wisconsin Department of Natural Resources, Bureau of Research.

Kaup, Katherine Palmer. 2000. *Creating the Zhuang: Ethnic Politics in China.* Boulder, CO: Lynne Rienner Publishers.

Khan, Shakil, and Ai Sugie. 2015. "Sand Mining and Its Social Impacts on Local Society in Rural Bangladesh: A Case Study of a Village in Tangail District." *Journal of Urban and Regional Studies on Contemporary India* 2(1): 1–11. http://doi.org/10.15027/41500 (accessed September 12, 2017).

Kinsley, D. 1995. *Ecology and Religion: Ecological Spirituality in a Cross-Cultural Perspective.* Englewood Cliffs, NJ: Prentice Hall.

Kioko, E. M. 2016. "Turning Conflict into Coexistence: Cross-Cutting Ties and Institutions in the Agro-Pastoral Borderlands of Lake Naivasha Basin, Kenya." PhD diss., University of Cologne, Germany. http://kups.ub.uni-koeln.de/7064/ (accessed October 21, 2017).

Kirsch, Stuart. 2001. "Lost Worlds: Environmental Disaster, 'Culture Loss,' and the Law." *Current Anthropology* 42: 167–198.

Klingeman, P. C., R. L. Beschta, P. D. Komer, and J. B. Bradley. 1999. *Gravel-Bed Rivers in the Environment.* Highlands Ranch, CO: Water Resources Publications, LLC.

Koirala, Madhav Prasad, and Er. Buddhi Raj Joshi. 2017. "Construction Sand, Quality and Supply Management in Infrastructure Project." *International Journal of Advances in Engineering & Scientific Research* 4(4): 1–15.

Kondolf, G. M. 1993. "The Reclamation Concept in Regulation of Gravel Mining in California." *Journal of Environmental Planning and Management* 36: 397–409.

Kondolf, G. M. 1994a. "Geomorphic and Environmental Effects of Instream Gravel Mining." *Landscape and Urban Planning* 28: 225–243. https://doi.org/10.1016/0169-2046(94)90010-8 (accessed October 28, 2017).

Kondolf, G. M. 1994b. "Environmental Planning in Regulation and Management of Instream Gravel Mining in California." *Landscape and Urban Planning* 29: 185–199.

Kondolf, G. M. 1995a. "Five Elements for Effective Evaluation of Stream Restoration." *Restoration Ecology* 3(2): 133–136.

Kondolf, G. M. 1995b. "Geomorphological Stream Channel Classification in Aquatic Habitat Restoration: Uses and Limitations." *Aquatic Conservation Marine and Freshwater Ecosystems* 5(2): 127–141.

Kondolf, G. M. 1997. "Hungry Water: Effects of Dams and Gravel Mining on River Channels." *Environmental Management* 21: 533–551.

Kondolf, G. M. 1998a. "Environmental Effects of Aggregate Extraction from River Channels and Floodplains." In *Aggregate Resources: A Global Perspective,* edited by P. T. Bobrowsky, 113–129. Brookfield: A. A. Balkema.

Kondolf, G. M. 1998b. "Large Scale Extraction of Alluvial Deposits from Rivers in California: Geomorphic Effects and Regulatory Strategies." In *Gravel Bed Rivers in the Environment,* edited by P. C. Klingeman, R. L. Beschta, P. D. Komar and J. B. Bradley, 455–470. Colorado: Water Resources Publications.

Kondolf, G. M., M. Smeltzer, and L. Kimball. 2002. "Freshwater Gravel Mining and Dredging Issues." Report prepared for Washington Department of Fish and Wildlife, Washington Department of Ecology and Washington Department of Transportation, 122 pp.

Kostka, G., and W. Hobbs. 2013. "Embedded Interests and the Managerial Local State: The Political Economy of Methanol Fuel-Switching in China." *Journal of Contemporary China* 22: 204–218.

Kostka, Genia, and Niklas Anzinger. 2016. "Large Infrastructure Projects in Germany: A Cross-sectoral Analysis." In *Large Infrastructure Projects in Germany: Between*

Ambition and Realities, edited by Genia Kostka and Jobst Fiedler, 15–38. Berlin: Hertie School of Governance.

Kostka, Genia, and Jobst Fiedler, eds. 2016. *Large Infrastructure Projects in Germany: Between Ambition and Realities.* Cham, Switzerland: Palgrave Macmillan.

Kostka, Genia, and Arthur P. J. Mol. 2013. "Implementation and Participation in China's Local Environmental Politics: Challenges and Innovations." *Journal of Environmental Policy and Planning* 15: 3–16.

Kostka, Genia, and Jonas Nahm. 2017. "Central-Local Relations: Recentralization and Environmental Governance in China." *The China Quarterly* 231: 567–582.

Krause, C., M. Diesing, and G. Arlt. 2010. "The Physical and Biological Impact of Sand Extraction: A Case Study of the Western Baltic Sea." *Journal of Coastal Research*, special issue 51: 215–226.

Krause, F. 2010. "Thinking Like a River: An Anthropology of Water and Its Uses along the Kemi River, Northern Finland." PhD diss., University of Aberdeen, Scotland.

Krausmann, F., S. Gingrich, N. Eisenmenger, K.-H. Erb, H. Haber, and M. Fischer-Kowalski. 2009. "Growth in Global Materials Use, GDP and Population during the 20th Century." *Ecological Economics* 68: 2696–2705.

Krishnakumar, A. 2002. "Enviromnental Degradation of Two River Basins of Southem Kerala." PhD diss., University of Kerala, Thituvananthapuram, India.

Kull, C., and H. Rangan. 2015. "The Political Ecology of Weeds: A Scalar Approach to Landscape Transformations." In *The International Handbook of Political Ecology*, edited by Raymond L. Bryant, 487–500. Cheltenham, UK: Edward Elgar Publishing.

Ladson, A. R., and D. A. Judd. 2014. "A Review of the Effect of Floodplain Gravel Mining on River Stability." In *Proceedings of the 7th Australian Stream Management Conference*, edited by G. Vietz, I. Rutherfurd, and R. Hughes, 249–259. Townsville, Queensland: River Basin Management Society.

Lai, Hongyi. 2003. "The Religious Revival in China." *Copenhagen Journal of Asian Studies* 18: 40–64.

Landry, Pierre F. 2008. *Decentralized Authoritarianism in China.* New York: Cambridge University Press.

Lary, Diana. 1972. "Communism and Ethnic Revolt: Some Notes on the Chuang (Zhuang) Peasant Movement in Kwangsi, 1921–31." *The China Quarterly* 49: 126–135.

Lave, Rebecca. 2015. "Reassembling the Structural: Political Ecology and Actor-Network Theory." In *The Routledge Handbook of Political Ecology*, edited by Tom Perreault, Gavin Bridge, and James McCarthy, 213–223. New York: Routledge.

Lei, Kuang, and Liu Li. 2012. "Discrimination against Rural-to-Urban Migrants: The Role of the Hukou System in China." *PLoS One.* November 5. https://doi.org/10.1371/journal.pone.0046932.

Lertzman, David A., and Harrie Vredenburg. 2005. "Indigenous Peoples, Resource Extraction and Sustainable Development: An Ethical Approach." *Journal of Business Ethics* 56: 239–254.

Lewis, J. 2015. *Green Innovation in China: China's Wind Power Industry and the Global Transition to a Low-Carbon Economy.* New York: Columbia University Press.

LGCXT (Xintang Town People's Government 新棠镇人民政府). 2010–2015. *Xintangzhen difangzhi* 新棠镇地方志 [Local Government Chronicles of Xintang Town]. Xintang Town: Xintang Town People's Government.

Li, Chao 黎超. 2016. "Guangxi chutai shoubu hedao caisha guanli tiaoli daocai zuigao fa 50 wan" 广西出台首部河道采砂管理条例 盗采最高罚50万 [Guangxi Promulgated the First River Sand Mining Management Regulations, Maximum Fine of 500,000 Yuan]. http://www.gxnews.com.cn/staticpages/20161201/newgx5840451a-15738872.shtml (accessed August 21, 2018).

Li, Hengchang 李恒昌. 2006. " Guangxi hedao caisha guifanhua guanli tantao" 广西河道采砂规范化管理探讨 [Discussion on Standardized Management of Sand Mining in Guangxi River Channel]. *Renmin Zhujiang* 人民珠江 [Pearl River], 16–18.

Li, Keqiang. 2016. "Report on the Work of the Government." Delivered at the Fourth Session of the 12th National People's Congress of the People's Republic of China. http://www.npc.gov.cn/zgrdw/englishnpc/Speeches/2016-03/18/content_1985677.htm (accessed October 10, 2017).

Li, Keqiang. 2017. "Report on the Work of the Government." http://www.xinhuanet.com/english/china/2017-03/16/c_136134017.htm (accessed October 15, 2017).

Li, Keqiang. 2020. "Report on the Work of the Government." Beijing: Delivered at the Third Session of the 13th National People's Congress of the People's Republic of China. http://www.xinhuanet.com/english/2020-05/29/c_139099211.htm (accessed June 5, 2021).

Li, Rongxin 李荣梓, and Zuichun Pang 庞醉春. 2000. *Qinzhou shizhi* 钦州市志 [Gazetteers of Qinzhou City]. Qinzhou City: Guangxi People's Publishing House.

Li, Tania Murray. 2005. "Beyond 'the State' and Failed Schemes." *American Anthropologist* 107(3): 383–394.

Li, Tania Murray. 2014. "What is Land? Assembling a Resource for Global Investment." *Transactions of the Institute of British Geographers* 39(4): 589–602. https://doi.org/10.1111/tran.12065 (accessed June 5, 2017).

Li, Xi. 2011. "Public Rental Housing Development in China." Master of science thesis, Department of Real Estate and Construction Management, KTH Royal Institue of Technology, Stockholm, Sweden.

Li, Yaofeng 李耀锋. 2016. "Xiangmuzhi yanjiu de xianzhuang yu xin lujing" 项目制研究的现状与新路径 [The Status and New Path of Project System Research]. *Chongqing daxue xuebao (shehui kexue ban)* 重庆大学学报（社会科学版） [Journal of Chongqing University (Social Science Edition)] 5: 107–113.

Li, Yifei, and Judith Shapiro. 2020. *China Goes Green: Coercive Environmentalism for a Troubled Planet.* Cambridge: Polity Press.

Liang, Jinjie 梁进杰, ed. 2000. *Guangxi tongzhi shaoshu minzu yuyan zhi* 广西通志.少数民族语言志 [Guangxi Gazetteer: Local Chronicles of Minority Language]. Nanning: Guangxi People's Publishing House.

Liang, Jun. 2018. "China Int'l Garden Expo Opens in Nanning." *People's Daily Online.* December 7. http://en.people.cn/n3/2018/1207/c90000-9526151.html (accessed March 11, 2019).

Liang, Yu. 2018. "Xi Congratulates Chinese Farmers for First Harvest Festival." *Xinhuanet.* September 22. http://www.xinhuanet.com/english/2018-09/22/c_137486620.htm (accessed October 1, 2018).

Liao, Mingjun 廖明君. 2004. *Zhuangzu ziran chongbai wenhua* 壮族自然崇拜文化 [The Culture of Nature Worship among the Zhuang]. Nanning: Guangxi Renmin Publishing House.

Liao, Shao-Gang, Jia-Yi Zhong, and Xue-Man Fu. 2018. "Research on River Chiefs System in China: Effect Evaluation, Defects and Future Prospects." *International Journal of Natural Resource Ecology and Management* 3: 24–31.

Lieberthal, K. 1997. "China's Governing System and Its Impact on Environmental Policy Implementation." *China Environment Series* 1: 3–8.

Lin, Changhua 林昌华. 2014. "Quyu fazhan zhong chengzhenhua xiaolü yu jiuye jiegou dongyin—jiyu Fujiansheng de zhengshi" 区域发展中城镇化效率与就业结构动因—基于福建省的实证 [Analysis on the Relation between Efficiency of Urbanization and Employment Structure in Regional Economic Development: Based on the Empirical Research of Fujian]. *Chongqing ligong daxue xuebao* (*Shehui kexue ban*) 重庆理工大学学报（社会科学） Journal of Chongqing Institute of Technology (7): 48–54.

Ling, Chun-Hui 凌春辉. 2010. "Lun 'Mojing Buluoto' de zhuangzu shengtai lunli yiyun" 论《麼经布洛陀》的壮族生态伦理意蕴 [On the Ecological Ethics of the Zhuang People as Represented in "Mojing Buluoto"]. *Guangxi minzu xueyuan xuebao* (*zhexue shehui kexue ban*) 广西民族学院学报(哲学社会科学版) [Journal of Guangxi University for Nationalities (Philosophy and Social Sciences Edition)] 32(3): 90–94.

Lin, Fengchun 林凤春. 2007. "Guinan 'Tiao Lingtou' changben yanjiu" 桂南"跳岭头"唱本研究 [Study on the Scripture of Tiaolingtou in the South of Guangxi]. Masters thesis, Guangxi University, Nanning, China.

Liu, Jingling, Luan Yun, Su Liya. et al., 2010. Public participation in water resources management of Haihe River Basin, China: the analysis and evaluation of status quo. Procedia Environmental Science, Volume 2, p. 1750–1758.

Liu, Hui, Yongqin David Chen, Tao Liu, and Lu Lin. 2019. "The River Chief System and River Pollution Control in China: A Case Study of Foshan." *Water* 11(8): 1606.

Liu, Jiang-Guo, and J. Diamond. 2005. "China's Environment in a Globalizing World." *Nature* 435: 1179–1186.

Liu, Shiyang 刘诗洋. 2019. "Shahuang de mimi" 砂荒的秘密 [The Secret of the Sand Shortage]. http://www.infzm.com/contents/163141 (accessed February 22, 2020).

Liu, Zhiwei. 1995. "Lineage on the Sands: The Case of Shawan." In *Down to Earth: The Territorial Bond in South China*, edited by David Fauer and Helen F. Siu, 21–43. Stanford, CA: Stanford University Press.

Lora-Wainwright, A. 2013. *Fighting for Breath: Living Morally and Dying of Cancer in a Chinese Village.* Honolulu: University of Hawai'i Press.

Loughlin, John. 2018. "Territory and Governance." In *Handbook on the Geographies of Regions and Territories*, edited by Anssi Paasi, John Harrison, and Martin Jones, 197–207. Cheltenham, UK: Edward Elgar Publishing.

Lu, Nan 鲁南. 2017. "Liangdi jingfang chiqiang duizhi: nimen jiujing zai zheng shenme" 两地警方持枪对峙：你们究竟在争什么？ [Two Police Forces in Confrontation with Guns: What are You Fighting for?] http://club.qingdaonews.com/showAnnounce_2_5830328_1_0.htm (accessed December 30, 2017).

Luo, Ting. 2014. "Village Economic Autonomy and Authoritarian Control over Village Elections in China: Evidence from Rural Guangdong Province." PhD diss., The London School of Economics and Political Science, London, UK.

Ma, Yonghuan, Chuguo Wu, Hui Lin, Baorong Huang, and Liyang Su. 2018. "Insights for Conservation and Utilization: Reconstruction of China's Natural Resources Management System." *International Journal of Science, Technology and Society* 6: 52–62.

Mann, Michael. 1984. "The Autonomous Power of the State: Its Origins, Mechanisms and Results." *European Journal of Sociology / Archives Européennes de Sociologie* 25: 185–213.

Mao, Ye 毛野, and Cai-An Huang 黄才安. 2004. "Caisha dui hechuang bianxing yingxiang de shiyan yanjiu" 采沙对河床变形影响的试验研究 [Experimental Study on the Effects of Sand Mining on Riverbed Deformation]. *Shuili xuebao* 水利学报 [Journal of Hydraulic Engineering] 5: 64–69.

Marks, Robert B. 1996. "Commercialization without Capitalism. Processes of Environmental Change in South China, 1550–1850." *Environmental History* 1(1): 56–82.

Marks, Robert B. 1998. *Tigers, Rice, Silk, and Silt: Environment and Economy in Late Imperial South China.* Cambridge: University of Cambridge Press.

Marks, Robert B. 2012. *China: Its Environment and History. Lanham,* Md.: Rowman & Littlefield.

Marston, S. A. 2000. "The Social Construction of Scale." *Progress in Human Geography* 24(2): 2219–2242.

Maya, K., V. Santhosh, D. Padmalal, and S. R. Aneesh Kumar. 2012. "Impact of Mining and Quarrying in Muvattupuzha River Basin, Kerala—An Overview on Its Environmental Effects." *Bonfring International Journal of Industrial Engineering and Management Science* 2(SI 1): 36–40.

McCarthy, J. 2002. "First World Political Ecology: Lessons from the Wise Use Movement." *Environment and Planning A: Economy and Space* 34: 1281–1302.

Meador, Michael R., and April O. Layher. 1998. "Instream Sand and Gravel Mining: Environmental Issues and Regulatory Process in the United States." *Fisheries* 23: 6–13.

Mei, Ciqi, and Margaret M. Pearson. 2014. "Killing a Chicken to Scare the Monkeys? Deterrence Failure and Local Defiance in China." *The China Journal* 72: 75–97.

Meng, Jie. 2017. "Illegal Sand Mines Closed along Yangtze." Xinhhuanet. August 8. http://www.xinhuanet.com//english/2017-08/08/c_136508946.htm (accessed November 6, 2018).

MFAPRC. 2018. "The Zhuang Ethnic Minority." Website of the Ministry of Foreign Affairs of the People's Republic of China. https://www.fmprc.gov.cn/mfa_eng/ljzg_665465/3584_665493/t17927.shtml (accessed November 28, 2018).

MIIT (Ministry of Industry and Information Technology of the People's Republic of China 中华人民共和国工业和信息化部), NDRC (National Development and Reform Commission 中华人民共和国国家发展和改革委员会), MNR (Ministry of Natural Resources of the People's Republic of China 中华人民共和国自然资源部), MEE (Ministry of Ecology and Environment of the People's Republic of China 生态环境部), MOHURD (Ministry of Housing and Urban-Rural Development of the People's Republic of China 中华人民共和国住房和城乡建设部), MOT (Ministry of Transport of the People's Republic of China 中华人民共和国交通运输部), MWR (Ministry of Water Resources of the People's Republic of China 中华人民共和国水利部), MEM (Ministry of Emergency Management of the People's Republic of China 中华人民共和国应急管理部), SAMR (State Administration for Market Regulation 国家市场监督管理总局), and China-Railway (中国国家铁路集团有限公司). 2019. *Guanyu tuijin jizhi shashi hangye gaozhiliang fazhan de ruogan yijian* 关于推进机制砂石行业高质量发展的若干意见 [Some Opinions on Promoting the High-Quality Development of the Artificial Sand Industry]. Government Policy [政策], Beijing: Ministry of Industry and Information Technology [工信部联原] (2019) 239 http://www.gov.cn/zhengce/zhengceku/2019-11/13/content_5451478.htm (accessed June 17, 2021).

Mitchell, G. J. 2006. "Can Gravel Mining and Water Supply Wells Coexist?" Final report, Mitchell Centre for Environmental and Watershed Research. Orono: University of Maine.

MLRPRC. 2012. *2012 China Mineral Resources*. Government Report, Ministry of Land and Resources People's Republic of China, Beijing: Geological Publishing House.

MOHURD (Ministry of Housing and Urban-Rural Development of the People's Republic of China 中华人民共和国住房和城乡建设部). 2018. "Guanyu haisha zhili de zhuanxiang xingdong tongzhi (Jian zhi dian [2018] 22 hao)" 关于海砂治理的专项行动通知（建质电[2018]22号）[Notice on Launching Special Actions for Policing

Violations of Sea Sands were Jointly Launched by the Eight Ministries (agencies)]. http://www.greatchinaca.com/index.php/rule/info/325.html (accessed January 3, 2019).

Mol, Arthur. P. J., and Nell T. Carter. 2006. "China's Environmental Governance in Transition." *Environmental Politics* 15(2): 149–170. doi:10.1080/09644010600562765.

Montinola, Gabriella, Yingyi Qian, and Barry R. Weingast. 1995. "Federalism, Chinese Style: The Political Basis for Economic Success in China." *World Politics* 48(1): 50–81.

Mwesigye, Walter M. 2018. "Where Our Sand is Taken." *Daily Monitor*. March 6. https://www.monitor.co.ug/SpecialReports/Where-our-sand-taken/688342-4330906-f5ig3lz/index.html (accessed December 2, 2018).

MWR (Ministry of Water Resources of the People's Republic of China 中华人民共和国水利部). 2018. "Quanmian jianli hezhangzhi xinwen fabuhui" 全面建立河长制新闻发布会 [Press Conference on the Comprehensive Establishment of the River Chief System]. July 17, http://www.mwr.gov.cn/hd/zxft/zxzb/fbh20180717/ (accessed June 16, 2021).

MWR and MTPRC (Ministry of Transport of the People's Republic of China 中华人民共和国交通运输部). 2016. "Shuilibu jiaotong yunshubu lianhe daji feifa caisha baozhang Changjiang jingjidai fazhan zhanlue shunli shishi" 水利部交通运输部联合打击非法采砂保障长江经济带发展战略顺利实施 [MWR and MOT PRC Jointly Crack Down on Illegal Sand Mining to Ensure the Smooth Implementation of the Development Strategy of the Yangtze River Economic Belt]. February 4. http://www.mwr.gov.cn/xw/slyw/201702/t20170212_844475.html (accessed August 24, 2017).

MWR, MTPRC (Ministry of Transport of the People's Republic of China 中华人民共和国交通运输部), and MLR (Ministry of Land and Resources of the People's Republic of China 中华人民共和国国土资源部). 2015. "Guanyu jingyibu jiaqiang hedao caisha guanli gongzuo de tongzhi" 关于进一步加强河道采砂管理工作的通知 [Further Strengthening of the River Sand Management Notice]. August 7. http://www.mwr.gov.cn/zw/tzgg/tzgs/201702/t20170213_858612.html (accessed September 12, 2017).

MWRPRC (Ministry of Water Resources of the People's Republic of China 中华人民共和国水利部). 2018a. "Guanyu kaizhan quanguo hehu caisha zhuanxiang zhengzhi xingdong de tongzhi" 关于开展全国河湖采砂专项整治行动的通知 [Notice of the General Office of the Ministry of Water Resources on Launching the National River and Lake Sand Mining Special Remediation Action]. Beijing: Ministry of Water Resources of the People's Republic of China.

MWRPRC (Ministry of Water Resources of the People's Republic of China 中华人民共和国水利部). 2018b. "Shuilibu guanyu jiakuai tuijing hehu guanli huading gongzuo de tongzhi" 水利部关于加快推进河湖管理范围划定工作的通知 [Notice of the Ministry of Water Resources on Accelerating the Delimitation of the Management Scope of Rivers and Lakes]. December 20. http://www.mwr.gov.cn/zw/tzgg/tzgs/201812/t20181220_1057631.html (accessed December 29, 2018).

Myšička, Stanislav. 2009. "Decentralization and Authoritarianism in China." *Journal of Contemporary European Studies* 2: 61–66.

NBS (National Bureau of Statistics). 2001. *China Statistical Yearbook 2001*. Website of the National Bureau of Statistics, the People's Republic of China. Beijing: China Statistics Press. http://www.stats.gov.cn/english/statisticaldata/yearlydata/YB2001e/ml/indexE.htm.

NBS. 2019. "National Bureau of Statistics of China: Announcement of the National Bureau of Statistics on the Final Verification of GDP in 2017." Website of the National Bureau of Statistics, the People's Republic of China. http://www.stats.gov.cn/tjsj/zxfb/201901/t20190118_1645555.html (accessed February 20, 2019).

NDRC (National Development and Reform Commission). 2016. "Notice of the National Development and Reform Commission on Issuing the Guiding Rules for Implementing Public-Private Partnership Projects in Traditional Infrastructure Fields." *Law Info China*. http://lawinfochina.com/display.aspx?id=22845&lib=law (accessed March 20, 2019).

NEAC (National Ethnic Affairs Commission of the People's Republic of China 国家民族事务委员会). 2018. "*Buluotuo*" 布洛陀. https://www.neac.gov.cn/seac/mzwh/201807/1084171.shtml(accessed September 20, 2018).

Neumann, R. 2015. "Political Ecology of Scale." In *The International Handbook of Political Ecology*, edited by Raymond L. Bryant, 475–486. Cheltenham, UK: Edward Elgar Publishing.

Nine, Cara. 2018. "Territorial Rights and Justice." In *Handbook on the Geographies of Regions and Territories*, edited by Anssi Paasi, John Harrison, and Martin Jones, 208–218. Stanford, CA: Edward Elgar Publishing.

Norman, D. K., C. F. Cederholm, and W. S. Lingley. 1998. "Floodplains, Salmon Habitat and Sand and Gravel Mining." *Washington Geology* 26: 21–28.

NPCPRC79 (National People's Congress 全国人民代表大会). 1979. *The Criminal Law of the People's Republic of China*. Beijing: National People's Congress. http://english.mofcom.gov.cn/aarticle/lawsdata/chineselaw/200211/20021100050381.html (accessed June 17, 2021).

NPCPRC86 (National People's Congress 全国人民代表大会). 1986. *The Mineral Resources Law of the People's Republic of China*. Beijing: National People's Congress. http://english.mofcom.gov.cn/aarticle/lawsdata/chineselaw/200211/20021100053795.html (accessed June 17, 2021).

NPCPRC88 (National People's Congress 全国人民代表大会). 1988. *Water Law of the People's Republic of China*. Beijing: The Sixth National People's Republic of China. http://english.mofcom.gov.cn/article/lawsdata/chineselaw/200211/20021100053796.shtml (accessed June 17, 2021).

NPCPRC96 (National People's Congress 全国人民代表大会). 1996. *Mineral Resources Law of the People's Republic of China*. Beijing: National People's Congress. http://www.lawinfochina.com/display.aspx?id=1033&lib=law (accessed June 17, 2021).

NPCPRC97 (National People's Congress 全国人民代表大会). 1997. *The Flood Control Law of the People's Republic of China.* Beijing: The National People's Congress of the People's Republic of China. http://www.npc.gov.cn/zgrdw/englishnpc/Law/2007-12/11/content_1383581.htm (accesed June 17, 2021).

NPCPRC07 (National People's Congress 全国人民代表大会). 2007. *Property Law of the People's Republic of China.* Beijing: National People's Congress. http://www.npc.gov.cn/zgrdw/englishnpc/Law/2009-02/20/content_1471118.htm (accessed June 16, 2021).

O'Brien, Kevin J., and Lianjiang Li. 2006. *Rightful Resistance: Contentious Politics in Rural China.* New York: Cambridge University Press.

GOCCPCC (General Office of the CCP Central Committee 中共中央办公厅) and GOSC (General Office of the State Council 国务院办公厅). 2017. "Guanyu quanmian tuixing hezhangzhi de yijian" 关于全面推行河长制的意见 [Office of the Central Committee of the Communist Party of China and Office of the State Council Issued the "Opinions on the Full Implementation of the River Chef System"]. http://www.gov.cn/gongbao/content/2017/content_5156731.htm (accessed December 20, 2017).

OLVCPRC10 (National People's Congress 全国人民代表大会). 2010. *The Organic Law of the Village Committees of the People's Republic of China.* Beijing: National People's Congress of China.

OLVCPRC87 (National People's Congress 全国人民代表大会). 1987. *The Organic Law of the Village Committees of the People's Republic of China.* Beijing: National People's Congress of China. http://english.mofcom.gov.cn/aarticle/lawsdata/chineselaw/200211/20021100050374.html (accessed June 17, 2021).

OLVCPRC98 (National People's Congress 全国人民代表大会). 1998. *The Organic Law of the Village Committees of the People's Republic of China.* Beijing: National People's Congress of China. http://www.npc.gov.cn/zgrdw/englishnpc/Law/2007-12/11/content_1383542.htm (accessd June 17, 2021).

Ostrom, Elinor. 1990. *Governing the Commons: The Evolution of Institutions for Collective Action.* Cambridge: Cambridge University Press.

Ostrom, Elinor. 2001. "Environment and Common Property Institutions." In *International Encyclopedia of the Social & Behavioral Sciences*, edited by Neil J. Smelser and Paul B. Baltes, 4560–4566. https://doi.org/10.1016/B0-08-043076-7/04146-2 (accessed June 17, 2017).

Padmalal, D., K. Maya, S. Sreebha, and P. R. Arun. 2005. "Impacts of River Sand Mining: An Overview of Kerala Rivers." *Proceedings of Workshop-Cum-Training on River Management, Centre for Earth Science Studies.* Trivandrum: Dept. of Geology, University of Kerala.

Padmalal, D., K. Maya, S. Sreebha, and R. Sreeja. 2008. "Environmental Effects of River Sand Mining: A Case from the River Catchments of Vembanad Lake, Southwest Coast of India." *Environmental Geology* 54: 879–889.

Paper, J. 2001. " 'Daoism' and 'Deep Ecology' Fantasy and Potentiality." In *Daoism and Ecology: Ways Within a Cosmic Landscape*, edited by N. J. Girardot, J. Miller, and X.-G. Liu. Cambridge, MA: Center for the Study of World Religions, Harvard Divinity School.

Paterson-Jones, Noel Gregor. 2019. "Tackling Europe's Infrastructure Needs: The European Fund for Strategic Investments." *Journal of Mega Infrastructure & Sustainable Development* 1(1): 54–78.

Paukert, C., J. Schloesser, J. Fischer, J. Eitzmann, K. Pitts, and D. Thornbrugh. 2008. "Effect of Instream Sand Dredging on Fish Communities in the Kansas River USA: Current and Historical Perspectives." *Journal of Freshwater Ecology* 23(4): 623–633.

Peet, Richard, and Michael Watts. 1996. *Liberation Ecologies: Environment, Development, Social Movements.* London: Routledge.

Pei, Xiaolin. 2005. "Collective landownership and its role in rural industrialization". In *Developmental Dilemmas : Land Reform and Institutional Change in China,* edited by P. Ho. London: Routledge: 203-227.

Pei, Shengji. 1985. "Some Effects of the Dai People's Cultural Beliefs and Practices upon the Environment of Xishuangbanna, Yunnan, China." In *Cultural Values and Human Ecology in Southeast Asia*, edited by Karl L. Huterer, A. Terry Rambo, and George Lovelace, 321–339. Ann Arbor: University of Michigan Press.

Pei, Shengji. 1993. "Managing for Biological Diversity Conservation in Temple Yards and Holy Hills: The Traditional Practices of the Xishuangbanna Dai Community, Southwest China." In *Ethics, Religion and Biodiversity: Relations between Conservation and Cultural Values*, edited by L. D. Hamilton, 118–132. Birmingham, UK: White Horses.

Peng, Bo 彭波, and Lei Mao 毛磊. 2014. "Huanbaofa ershiwu nian shouci daxiu" 环保法25年首次大修 [The First Major Overhaul of the Environmental Protection Law in 25 Years]. April 25. http://npc.people.com.cn/n/2014/0425/c14576-24940197.html (accessed April 3, 2016).

Pia, Andrea E. 2016. " 'We Follow the Reason, Not the Law': Disavowing the Law in Rural China." *Political and Legal Anthropology Review* 39(2): 276–293.

Pia, Andrea E. 2019. "Memory Leaks: Local Histories of Cooperation as a Solution to Water-related Cooperation Problems." In *Cooperation in Chinese Communities: Morality and Practice*, edited by Charles Stafford, Ellen R. Judd, and Eona Bell, 101–120. London: Bloomsbury Academic.

Prudham, Scott. 2015. "Property and Commodification." In *The Routledge Handbook of Political Ecology*, edited by Tom Perreault, Gavin Bridge, and James McCarthy, 430–445. London: Routledge.

Qi, Ye, and Lingyun Zhang. 2014. "Local Environmental Enforcement Constrained by Central-Local Relations in China." *Environmental Policy and Governance* 24(3): 216–232.

Qin, Cailian 覃彩銮. 1999. "Shilun Zhuangzu wenhua de ziran shengtai huanjing" 试论壮族文化的自然生态环境 [On the Natural Ecological Environment of Zhuang Culture]. *Xueshu luntan* 学术论坛 [Academic Forum] 6:115–119.

Qin, Cailian 覃彩銮. 1998. *Zhuangzu ganlan wenhua* 壮族干栏文化 [The Pile-dwelling Cultural of the Zhuang]. Guangxi: Guangxi National Publishing House [广西民族出版社].

Qin, 覃乃昌. 1997. *Zhuangzu daozuo nongyeshi* 壮族稻作农业史 [History of Rice Farming among the Zhuang]. Guangxi: Guangxi National Publishing House.

Qinzhou Water Conservancy Bureau Office. 2017. "Guanyu chengli Qinzhoushi tuixing hezhangzhi gongzuo lingdao xiaozu de tongzhi" 关于成立钦州市推行河长制工作领导小组的通知 [Notice on the Establishment of the Leading Group for the Implementation of the River Long System in Qinzhou City]. http://www.qzwater.gov.cn/Article.aspx?sid=1&id=12565 (accessed October 12, 2018).

Qinzhou360. 2017. "A Big Occasion: The District of Qinnan Dispatched 20 Units to Rectify Illegal Sand-Selling Sites." http://bbs.qinzhou360.com/thread-2415017-1-1.html (accessed November 11, 2017).

QLCCC (Qinzhou Local Chronicles Compilation Committee Office 钦州地方志编纂委员会办公室). 2012. *Qinzhou shixianzhen gaikuang* 钦州市县镇概况 [An Overview of the Counties and Towns of Qinzhou City]. Guangxi Fine Arts Publishing House.

QLCCC. 2018. *Qinzhou shizhi* 钦州市志 [Chronicles of Qinzhou City]. Qinzhou City: Guangxi People's Publishing House.

Qu, Jingdong. 2012. "The Project System: A New Form of State Governance." *Social Sciences in China* 33(4): 113–130.

QZGOV (Qinzhou Municipal People's Government 钦州市人民政府). 2008. "Guangxi Zhuangzu zizhiqu Qinzhoushi renmin zhengfu bangongshi guanyu yinfa Qinjiang Maolingjiang Dafengjiang bufen heduan hedao caishaquan churang fangan" 广西壮族自治区钦州市人民政府办公室关于印发《钦江茅岭江大风江部分河段河道采砂权出让方案》的通知 [Notice of the Office of Qinzhou Municipal People's Government of Guangxi Zhuang Autonomous Region on Printing and Distributing the Scheme for Transferring Sand Mining Rights on Parts of the Maoling River and Dafeng River of Qinjiang River]. http://m.law-lib.com/law/law_view.asp?id=256943&page=1 (accessed July 14, 2017).

QZGOV (Qinzhou Municipal People's Government 钦州市人民政府). 2017. "Min zu" 民族 [The Minority]. http://www.qinzhou.gov.cn/glqz_205/mz/ (accessed August 30, 2018).

Rappaport, R. A. 1967. "Ritual Regulation of Environmental Relations among a New Guinea People." *Ethnology* 6(1): 17–30.

Rappaport, R. A. 1968. *Pigs for the Ancestors.* New Haven: Yale University Press.

Ran, Ran. 2013. "Perverse Incentive Structure and Policy Implementation Gap in China's Local Environmental Politics." *Journal of Environmental Policy & Planning* 15:1, 17–39.

RCOQ (Qinzhou Water Conservancy Bureau 钦州水利局). 2017. "Qinzhoushi hezhangzhi bangongshi guanyu luoshi zizhiqu ducha Qinzhoushi quanmian tuixing hezhangzhi gongzuo fankui yijian zhenggai qingkuang de baogao" 钦州市河长制办公室关于落实自治区督查钦州市全面推行河长制工作反馈意见整改情况的报告 [Report on the Rectification of the Feedback on the Implementation of the River Chief System in the Autonomous Region in Qinzhou City]. Qinzhou Water Conservancy Bureau. http://www.qzwater.gov.cn/article.aspx?id=14769&cid=1065 (accessed October 28, 2018).

Robbins, Paul. 2004. *Political Ecology: A Critical Introduction.* Oxford: Blackwell.

Robbins, Paul. 2012. *Political Ecology: A Critical Introduction.* 2nd ed. Oxford: Wiley-Blackwell.

Rockefeller, S. C., and J. C. Elder, eds. 1992. *Spirit and Nature: Why the Environment is a Religious Issue*. Boston: Beacon Press.

Ruoxiu, Ouyang 欧阳若修. 1986. *Zhuangzu wenxue shi* 壮族文学史 [History of Zhuang Literature]. Vol. 1. Nanning: Guangxi Renmin Publishing House.

Safi, Michael. 2017. "Villagers Pay Tragic Price as Indian Building Boom Drives Demand for Sand." *The Guardian*. December 30. https://www.theguardian.com/world/2017/dec/30/india-sand-mining-conflict-deaths-building-boom-environmental-damage (accessed June 5, 2018).

Sayer, Andrew. 2000. "Moral Economy and Political Economy." *Political Economy* 61: 79–104.

Sayer, Andrew. 2004. *Moral Economy.* Lancaster University: Department of Sociology. https://www.lancaster.ac.uk/fass/resources/sociology-online-papers/papers/sayer-moral-economy.pdf (accessed April 5, 2017).

Scott, J. C., 1985. *Weapons of the Weak: Everyday Forms of Peasant Resistance*. USA: Yale University Press.

Scott, J. C., 1990. *Domination and the Arts of Resistance: Hidden Transcripts*. [New Haven, CT]:Yale University Press.

Scott, James C. 2012. "Infrapolitics and Mobilizations: A Response by James C. Scott." *Revue française d'études américaines*, no. 131: 112–117.

SCPCGZAR (Standing Committee of the People's Congress of Guangxi Zhuang Autonomous Region 广西壮族自治区人民代表大会常务委员会). 2016. "Guangxi Zhuangzu zizhi qu hedao caisha guanli tiaoli" 广西壮族自治区河道采砂管理条例 [Guangxi Zhuang Autonomous Region River Sand Mining Management Regulations]. http://www.gxzf.gov.cn/zwgk/flfg/dfxfg/20161222-634500.shtml (accessed July 11, 2018).

SCPRC87 (Order of the State Council of the People's Republic of China 中华人民共和国国务院令). 1987. *Zhonghua renmin gongheguo hangdao guanli tiaoli* 中华人民共和国航道管理条例 [The Regulations of the People's Republic of China on the Administration of the Waterways of the People's Republic of China]. Beijing: State Council of the People's Republic of China.

SCPRC88 (Order of the State Council of the People's Republic of China 中华人民共和国国务院令). 1988. *Zhonghua renmin gongheguo hedao guanli tiaoli* 中华人民共和国河道管理条例 [The Regulations of the People's Republic of China on River Administration]. Beijing: State Council of the People's Republic of China.

SCPRC94 (Order of the State Council of the People's Republic of China 中华人民共和国国务院令). 1994. *Zhonghua renmin gongheguo kuangchan ziyuan fa shishi xize* 中华人民共和国矿产资源法实施细则 [The Rules for the Implementation of the Mineral Resources Law of the People's Republic of China]. Beijing: State Council of the People's Republic of China.

Sekhar, L. K., and S. K. Jayadev. 2003. "Karimanal, Mineral Beach-Sand, Mining in the Alappuzha Coast of Kerala: A People's Perspective." In *Proceedings of the Third International Conference on Environment and Health (ICEH): Urban Planning and Environmental Management for Human Health, Chennai, India, December 15–17 2003*, edited by M. J. Bunch, V. Madha Suresh, and T. Vasantha Kumara, 470–488. Chennai, India: Department of Geography, University of Madras.

Senanayake, S. G. J. N. 2006. "Indigenous Knowledge as a Key to Sustainable Development." *Journal of Agricultural Sciences—Sri Lanka* 2(1): 87–94.

Shapiro, Judith. 2001. *Mao's War against Nature: Politics and the Environment in Revolutionary China*. Cambridge: Cambridge University Press.

Shapiro, Judith. 2012. *China's Environmental Challenges*. Cambridge: Polity Press.

Sheppard, E., and R. McMaster. 2004. "Introduction: Scale and Geographic Inquiry." In *Scale and Geographic Inquiry*, edited by E. Sheppard and R. McMaster, 1–22. Oxford: Blackwell.

Siu, Helen F. 1989a. "Recycling Rituals: Politics and Popular Culture in Contemporary Rural China." In *Unofficial China: Popular Culture and Thought in the People's Republic*, edited by Perry Link, Richard Madsen, and Paul Pickowicz, 121–137. Canada: HarperCollins Canada.

Siu, Helen F. 1989b. "Socialist Peddlers and Princes in a Chinese Market Town." *American Ethnologist* 16: 195–212.

Siu, Helen F. 2016. *Tracing China: A Forty-Year Ethnographic Journey*. Hong Kong: Hong Kong University Press.

Skinner, George William. 1977. "Cities and the Hierarchy of Local System." In *The City in Late Imperial China*, edited by G. William Skinner, 275–352. Stanford, CA: Stanford University Press.

Smil, Vaclav. 1997. "China's Environment and Security: Simple Myths and Complex Realities." *SAIS Review* 17(1): 107–126.

Smith, N. 2000. "Scale." In *The Dictionary of Human Geography*, edited by R. J. Johnston, D. Gregory, G. Pratt, and M. Watts, 724–727. Malden, MA: Blackwell.

Smith, N. 2007. "Nature as an Accumulation Strategy" In *Socialist Register 2007: Coming to Terms with Nature*, edited by Leo Panitch and Colin Leys, 16–36. Monmouth, UK: The Merlin Press.

Soman, K., and K. Baji. 1999. "Status of Sand Mining Sites in Kallada River, Kollam District." Centre for Earth Science Studies, Thiruvananthapuram.

Sreebha, S. 2008. "The Environmental Impact of Sand Mining: A Case Study in the River Catchments of Vembanad Lake, Southwest India." PhD diss., Cochin University of Science and Technology, Kochi, India.

Stern, Rachel E. 2013. *Environmental Litigation in China: A Study in Political Ambivalence*. Cambridge: Cambridge University Press.

Strang, V. 2014. "Fluid Consistencies: Material Relationality in Human Engagements." *Archaeological Dialogues* 21: 133–150.

Stronberg, Joel B. 2016. "True Grit: The Sand Mining Debate." August 8. https://www.linkedin.com/pulse/true-grit-sand-mining-debate-joel-b-stronberg (accessed January 3, 2018).

Sturgeon, Janet C. 2007. "Pathways of 'Indigenous knowledge' in Yunnan, China." *Alternatives: Global, Local, Political* 32(1): 129–153.

Subramanian, Samanth. 2017. "How Singapore is Creating More Land for Itself." *The New York Times*. April 20. https://www.nytimes.com/2017/04/20/magazine/how-singapore-is-creating-more-land-for-itself.html?mcubz=3&_r=0?xid=PS_smithsonian (accessed October 12, 2019).

Sullivan, S. 2013. "Banking Nature? The Spectacular Financialization of Environmental Conservation." *Antipode* 45: 198–217.

Suo, Youwei 索有为, Wanting Xi 奚婉婷, and Lu Zhang 张璐. 2012. "Guangdong chutai xingui fang hesha cheng gebie diqu fubai wenchuang" 广东出台新规防河砂开采成个别地区腐败温床 [Guangdong Introduced a New Regulation to Prevent River-Sand Mining from Becoming a Hotbed of Corruption in Individual Areas]. http://news.sohu.com/20120807/n350090261.shtml (accessed June 28, 2021).

Swyngedouw, Erik. 2015. "Depoliticized Environments and the Promises of the Anthropocene." In *The International Handbook of Political Ecology*, edited by Raymond L. Bryant, 131–146. Chelterham, UK: Edward Elgar Publishing.

Tan, Xingxing. 2015. "Guangxi Will Promote the Renovation of 260,000 Rural Dilapidated Houses This Year" [广西今年将推进26万户农村危房改造 户均补助1.5万元]. http://www.gov.cn/xinwen/2015-07/11/content_2895483.htm (accessed October 22, 2019).

Tilt, Bryan. 2007. "The Political Ecology of Pollution Enforcement in China: A Case from Sichuan's Rural Industrial Sector." *The China Quarterly* 192: 915–932.

Took, Jennifer. 2005. *A Native Chieftaincy in Southwest China: Franchising a Tai Chieftaincy under the Tusi System of Late Imperial China*. Leiden: Brill.

Torres, Aurora, Jodi Brandt, Kristen Lear, and Jianguo Liu. 2017. "A Looming Tragedy of the Sand Commons." *Science*, 357: 970–971. https://science.sciencemag.org/content/357/6355/970.full/ (accessed October 26, 2019).

Torres, Aurora, Jianguo "Jack" Liu, Jodi Brandt, and Kristen Lear. 2017. "The World is Running Out of Sand: The Little-Known Exploitation of This Seemingly Infinite Resource Could Wreak Political and Environmental Havoc." *Smithsonian*. September 8. https://www.smithsonianmag.com/science-nature/world-facing-global-sand-crisis-180964815/(accessed October 26, 2019).

Ulloa, Astrid. 2015. "Environment and Development: Reflections from Latin America." In *The Routledge Handbook of Political Ecology*, edited by Tom Perreault, Gavin Bridge, and James McCarthy, 320–331. London: Routledge.

UN (United Nations). 2015. "Who Are Indigenous Peoples?" Website of the United Nations. https://unstats.un.org/UNSD/geoinfo/UNGEGN/docs/26th-gegn-docs/UN%20PFII%205session_factsheet1.pdf (accessed May 20, 2021).

UNEP (United Nations Environment Programme). 2014. "Sand, Rarer than One Thinks." *UNEP Global Environmental Alert Service*. March 2014. https://wedocs.unep.org/bitstream/handle/20.500.11822/8665/GEAS_Mar2014_Sand_Mining.pdf?sequence=3 (accessed August 10, 2017).

USGS (United States Geological Survey). 2013a. *Cement, Statistics, and Information*. Reston, VA: U.S. Geological Survey, Mineral Commodity Summaries.

USGS. 2013b. "Sand and Gravel (Construction) Statistics." In *Historical Statistics for Mineral and Material Commodities in the United States*, edited by T. D. Kelly and G. R. Matos, 136–137. U.S. Geological Survey Data Series 140. Reston, VA: U.S. Geological Survey.

USGS. 2013c. *The 2011 Minerals Yearbook: Cement*. Reston, VA: U.S. Geological Survey.

Vaccaro, I., O. Beltran, and P. A. Paquet. 2013. "Political Ecology and Conservation Policies: Some Theoretical Genealogies." *Journal of Political Ecology* 20(1): 255–272.

Valdés-Rodríguez, O. A., and Arturo Pérez-Vázquez. 2011. "Sustainable Livelihood: An Analysis of the Methodology." *Tropical and Subtropical Agroecosystems* 14: 91–99.

Van Rooij, Benjamin. 2006a. "Implementation of Chinese Environmental Law: Regular Enforcement and Political Campaigns." *Development and Change* 37(1): 57–74.

Van Rooij, Benjamin. 2006b. *Regulating Land and Pollution in China: Lawmaking, Compliance, and Enforcement*. Leiden: Leiden University Press.

Vayda, Andrew. 1983. "Progressive Contextualization: Methods for Research in Human Ecology." *Human Ecology* 11: 265–281.

Viju, B. 2009. "Sand Mining Issue Haunts Naik." *Times of India*. October 27.

Von Glasenapp, Markus, and Thomas F. Thornton. 2011. "Traditional Ecological Knowledge of Swiss Alpine Farmers and Their Resilience to Socioecological Change." *Human Ecology* 39: 769–781.

Wagner, John R., and Jerry K. Jacka. 2018. *Island Rivers: Fresh Water and Place in Oceania*. Canberra, Australia: Australian National University Press.

Wagner, John R., Jerry K. Jacka, Edvard Hviding, Alexander Mawyer, and Marama Muru-Lanning. 2018. "Introduction: River as Ethnographic Subject." In *Fresh Water and Place in Oceania*, by John R. Wagner and Jerry K. Jacka. Canberra, Australia: The Australian National University Press.

Wallis, C. 2011. "New Media Practices in China: Youth Patterns, Processes, and Politics." *International Journal of Communication* 5: 406–436.

Walter, Mariana. 2014. "Political Ecology of Mining Conflicts in Latin America: Analysis of Environmental Justice Movements and Struggles over Scales." PhD diss., Institut de Ciència i Tecnologia Ambientals, Universitat Autònoma de Barcelona.

Wang, Alex L. 2013. "The Search for Sustainable Legitimacy: Environmental Law and Bureaucracy in China." *Harvard Environmental Law Review* 37(2): 365–440.

Wang, Alex L. 2018. "Symbolic Legitimacy and Chinese Environmental Reform." *Environmental Law* 48: 17–39.

Wang, Jin. 2010. "China's Green Laws are Useless." *China Dialogue.* September 23. https://chinadialogue.net/en/pollution/3831-china-s-green-laws-are-useless/ (accessed October 22, 2017).

Wang, Junwei 王军伟, and Hongming Xiong 熊红明. 2015. "Guangxi Qinzhou shuili woan luoma guanyuan: wo meiquan, laoban que pinming saiqian" 广西钦州水利窝案落马官员：我没权，老板却拼命塞钱 [Guangxi Qinzhou Water Conservancy Corruption. Government Official: I Have No Power, But the Boss Desperately Sends Money]. https://www.guancha.cn/politics/2015_05_21_320384.shtml (accessed October 12, 2017).

Wang, Linzhu. 2015. "The Identification of Minorities in China." *Asian-Pacific Law and Policy Journal* 16(2): 1–21.

Wang, Mingfu 王明富. 2003. "Yunnansheng maguanxian A'e xinzhai ji Buluotuo shenshu diaocha" 云南省马关县阿峨新寨祭布洛陀神树调查 [Investigation of the Buluotuo God Tree in the A'e Village, Maguan County, Yunnan Province]. *Wenshan shifan gaodeng zhuanke xuexiao xuebao* 文山师范高等专科学校学报 [Wenshan Teachers' College and College Journal] 16(1): 13–15.

Wang, Na 王娜, and Bin Li 李斌. 2014. "Sahala shamo shagao qiangdu hunningtu peihebi sheji ji yanjiu" 撒哈拉沙漠砂高强度混凝土配合比设计及研究 [Design and Research on the Mixing Proportions of Sahara Desert Sand High-Strength Concrete]. *Concrete*, 139–142.

Wang, Zhu-Guang 王祝光. 1988. *Guangxi nongcun hezuo jingji shiliao II* 广西农村合作经济史料II [Historical Data of Rural Cooperative Economy in Guangxi II]. Guangxi: Guangxi People's Publishing House.

Watts, M.J. 1985. "Social theory and environmental degradation: the case of Sudano-Sahelian West Africa." In Desert development: man and technology in sparselands, edited by Y. Gradus, Dordrecht: D. Reidel:14–32.

Wegrich, Kai, Genia Kostka, and Gerhard Hammerschmid, eds. 2017. *The Governance of Infrastructure.* Oxford: Oxford University Press.

Weir, J. K. 2009. *Murray River Country: An Ecological Dialogue with Traditional Owners.* Canberra, Australia: Aboriginal Studies Press.

Wen, Jiabao 温家宝. 2009."Zhengfu gongzuo baogao" 政府工作报告 [Report on the Work of the Government]. Government Report, the Second Session of the Eleventh National People's Congress, Beijing: Premier of the State Council. http://www.qingdaonews.com/qxw/content/2009-03/16/content_8011594.htm (accessed September 29, 2018).

Wen, Jiabao 温家宝. 2012."Zhengfu gongzuo baogao" 政府工作报告 [Report on the Work of the Government]. Government Report, the Fifth Session of the Eleventh National People's Congress, Beijing: Premier of the State Council. http://theory.people.com.cn/n/2013/0319/c40531-20832391.html (accessed September 20, 2018).

White, B., S. M. Borras Jr, R. Hall, I. Scoones, and W. Wolford. 2012. "The New Enclosures: Critical Perspectives on Corporate Land Deals." *Journal of Peasant Studies* 39: 619–647.

White, Lynn, Jr. 1967. "The Historical Roots of Our Ecologic Crisis." *Science* 10: 1203–1207.

Whitmore, Andy, Geoff Nettleton, and Leah Abayao, eds. 2012. *Pitfalls and Pipelines: Indigenous Peoples and Extractive Industries.* Baguio City, Philippines: Tebtebba Foundation and IWGIA.

Williams, D. 2016. *Depoliticized Infrastructure Governance and the Case of Infrastructure Australia.* Berlin: Hertie School of Governance.

Willnat, Lars, Lu Wei, and Jason A. Martin. 2015. "Politics and Social Media in China." In *The Routledge Handbook of Chinese Media*, edited by Gary D. Rawnsley and Ming-yeh T. Rawnsley, 199–220. London: Routledge.

Wong, Edward. 2013. "Cost of Environmental Damage in China Growing Rapidly Amid Industrialization." *The New York Times.* March 29. https://www.nytimes.com/2013/03/30/world/asia/cost-of-environmental-degradation-in-china-is-growing.html#:~:text=Cost%20of%20Environmental%20Damage%20in%20China%20Growing%20Rapidly%20Amid%20Industrialization,-Workers%20demolishing%20the&text=BEIJING%20%E2%80%94%20The%20cost%20of%20environmental,news%20report%20said%20this%20week (accessed September 11, 2016).

Wu, Jing, and I-Shin Chang. 2020. *Environmental Management in China: Policies and Institutions.* Singapore: Springer.

Wu, Zhiguang 吴志广. 2008. *Changjiang hedao caisha guanli shijian yu tansuo* 长江河道采砂管理实践与探索 [Practice and Exploration of Sand Mining Management in the Yangtze River]. Wuhan: Changjiang Press.

Xi, Jinping. 2017. "Secure a Decisive Victory in Building a Moderately Prosperous Society in All Respects and Strive for the Great Success of Socialism with Chinese Characteristics for a New Era." *Xinhuanet.* October 18. http://www.xinhuanet.com/english/download/Xi_Jinping's_report_at_19th_CPC_National_Congress.pdf (accessed March 12, 2018).

Xi, Jinping. 2020. "The Yangtze River Protection Law of the People's Republic of China." Order of the President of the People's Republic of China, No 65. http://www.npc.gov.cn/englishnpc/c23934/202103/8cccd3c25caf464d8ae3aedc66a21dd0.shtml (accessed March 1, 2021).

Xia, Wei-Man 夏玮蔓. 2018. "Qinzhou jianli peitao zhidu quanmian tuixing hezhangzhi" 钦州建立配套制度全面推行河长制 [Qinzhou Set Up a Comprehensive System to Implement the River Chief System]. http://www.gxnews.com.cn/staticpages/20180104/newgx5a4da2cb-16811452.shtml (accessed January 10, 2018).

Xiao, Shasha, and Qianyu Tang. 2016. "Baoli qudong feifa daocai hesha, yifa yancheng daocai zhe yu 'Baohusan' shi guanjian" 暴利驱动非法盗采河砂 依法严惩盗采者与'保护伞'是关键 [Profiteering Drives the Illegal Exploitation of River Sand, According to Law to Punish Illegal Miners, and Its "Umbrella" is the Key Method]. http://www.xjrb.com/2016/0503/301530.shtml (accessed December 30, 2017).

Xie, Lei. 2015. "Political Participation and Environmental Movements in China." In *The International Handbook of Political Ecology*, edited by Raymond L. Bryant, 246–259. Chelterham, UK: Edward Elgar Publishing.

Xinhua News Agency. 2016. "The 13th Five-Year Plan for the National Economic and Social Development of the People's Republic of China." *Xinhuanet.* March 17. http://news.xinhuanet.com/politics/2016lh/2016-03/17/c_1118366322.htm (accessed January 10, 2018).

XTGS (Xintang Town People's Government 新棠镇人民政府). 2013–2015. *Xintangzhen xiangzhen nongye zongtaizhang* 新棠镇乡镇农业总台帐 [Xintang Township Government Statistics]. Xintang Town: Xintang Town People's Government.

Xu, Jianchu, Jefferson Fox, Xing Lu, Nancy Podger, Stephen Leisz, and Xishu Ai. 1999. "Effects of Swidden Cultivation, State Policies, and Customary Institutions of Land Cover in a Hani Village, Yunnan, China." *Mountain Research and Development* 19: 123–132.

Xu, Jianchu, Erzi T. Ma, Duojie Tashi, Yongshou Fu, Zhi Lu, and David Melick. 2005. "Integrating Sacred Knowledge for Conservation: Cultures and Landscapes in Southwest China." *Ecology and Society* 10(2). https://www.ecologyandsociety.org/vol10/iss2/art7/ (accessed December 10, 2018).

Xu, Jianchu 许建初, Shengji Pei 裴盛基, and Sanyang Chen 陈三阳. 1997. "Indigenous Swidden Agroecosystems in Mengsong Hani Community." in *Xishuangbanna lunxie nongye shengtai xitong de shengwu duoyangxing yanjiu lun* 西双版纳轮歇农业生态系统的生物多样性研究论 [Biodiversity in Swidden Agroecosystems in Xishuangbanna], edited by Jianchu Xu and Shengji Pei, 26–33. Kunming: Yunnan Education Press.

Xu, Xiao-Ming 许晓明. 2006. "Jin shinian Zhuangzu minjian xinyang yanjiu zongshu" 近十年壮族民间信仰研究综述 [A Summary of the Research on the Folk Beliefs of the Zhuang Nationality in the Past Ten Years]. *Guangxi minzu xueyuan xuebao (zhexue shehui kexue ban)* 广西民族学院学报(哲学社会科学版) [Journal of Guangxi University for Nationalities (Philosophy and Social Sciences Edition)] SI 1, 100–102.

Xu, Yu 徐玉. 2011. *Dawenhe caisha dui fanghong gongcheng he shengtai huanjing de yingxiang* 大汶河采砂对防洪工程和生态环境的影响 [Impact of Sand Mining in the Dawen River on Flood Control Projects and the Ecological Environment]. Jinan: University of Jinan.

Yamineva, Yulia, and Zhe Liu. 2019. "Cleaning the Air, Protecting the Climate: Policy, Legal and Institutional Nexus to Reduce Black Carbon Emissions in China." *Environmental Science & Policy* 95: 1–10.

Yang, Chen 杨陈, Jianling Huang 黄剑铃, and Wei Wei 韦威. 2016. "Di shierjie Zhongguo (Nanning) guoji yuanbohui qidong jianshe, 2018 nian dong kaimu" 第十二届中国(南宁)国际园博会启动建设 2018 年冬开幕 [The Construction of the Twelfth China International Garden EXPO Starts and Will Open in Winter 2018]. https://web.archive.org/web/20180103121213/http://www.gx.chinanews.com/news/2016/1228/8243.html (accessed December 28, 2016).

Yang, Fenggang. 2011. *Religion in China: Survival and Revival under Communist Rule.* Oxford: Oxford University Press.

Yang, G. B. 2009. *The Power of the Internet in China: Citizen Activism Online.* New York: Columbia University Press.

Yang, Kun 杨堃. 1986. "Guanyu minzu he minzu wenti de jidian yijian" 关于民族和民族问题的几点意见 [Some Opinions on Nationality and Nationality Problems]. *Minzu yanjiu* 民族研究 [Ethno-national Studies] 4: 11–12.

Yang, Mayfair. 1996. "Tradition, Travelling Anthropology, and the Discourse of Modernity in China." In *The Future of Anthropological Knowledge*, edited by Henrietta L. Moore, 93–113. London: Routledge.

Yeh, Emily T. 2000. "Forest Claims, Conflicts and Commodification: The Political Ecology of Tibetan Mushroom-Harvesting Villages in Yunnan Province, China." *The China Quarterly* 161: 264–278.

Yeh, Emily T. 2007. "Tibetan Indigeneity: Translations, Resemblances, and Uptake." In *Indigenous Experience Today*, edited by Marisol De la Cadena and Orin Starn, 69–98. New York: Berg.

Yeh, Emily T. 2015. "Political Ecology in and of China." In *The International Handbook of Political Ecology*, edited by Raymond L. Bryant, 619–632. Cheltenham, UK: Edward Elgar Publishing.

Yin, Lun. 2015. *Climate Anthropology.* Beijing: Intellectual Property Publishing House.

Yin, Lun, Misiani Zachary, Yanyan Zheng, Xiaohan Zhang, and Antonine Sakwa. 2020. "Traditional Ecological Knowledge of Shifting Agriculture of Bulang People in Yunnan, China." *American Journal of Environmental Protection* 9: 56–63.

Yin, Lun, and Xiaohan Zhang. 2020. "Traditional Knowledge Is the Light of Wisdom for Conserving Biodiversity and Adapting to Climate Change." *Das Questöes* 10: 104–108.

Yü, Dan Smyer. 2014. "Sentience of the Earth: Eco-Buddhist Mandalizing of Dwelling Place in Amdo, Tibet." *Journal for the Study of Religion, Nature, and Culture* 8: 483–501.

Yü, Dan Smyer. 2017. "An Indigenous Cosmovisionary Turn in the Study of Religion and Ecology." In *The Routledge Handbook of Religion and Ecology*, edited by Willis Jenkins, Mary Evelyn Tucker, and John Gerim, 120–128. London: Routledge.

Yu, Xiaogang. 1993. "Protected Areas, Traditional Natural Resource Management Systems and Indigenous Women: Case Study in Xishuangbanna, PR China." Masters thesis, Asian Institute of Technology, Bangkok, Thailand.

Yu, Zilong 余子龙. 2014. "Jiyu DMSP/OLS yejing dengguang shuju de chengshi kongjian kuozhang tezheng yu qudong jizhi yanjiu" 基于 DMSP/OLS 夜景灯光数据的城市空间扩张特征与驱动机制研究 [Research of Urban Spatial Expansion and Driving Mechanism Based on DMSP/OLS Data—A Case Study of Guangxi]. Masters thesis, Fudan University, Shanghai, China.

Yuan, Jie-Feng 袁杰锋, and Zhen-Sheng Liu 刘震胜. 2000. "Changjiang hedao caisha guanli zhengce fagui yanjiu" 长江河道采沙管理政策法规研究 [Study on the Policy, Laws, and Regulations of Sand Dredging Management in the Changjiang River]. *Renmin Changjiang* 人民长江 [Yangtze River] 31(11): 22–24.

Zeng, Jie-Li 曾杰丽. 2008. "Zhuangzu minjian xinyang de hexie shengtai lunli yiyun" 壮族民间信仰的和谐生态伦理意蕴 [On the Ethics of Harmonious Ecology in the Folk Beliefs of the Zhuang People]. *Guangxi minzu daxue xuebao (zhexue shehui kexue ban)* 广西民族大学学报（哲学社会科学版） [Journal of Guangxi University for Nationalities (Philosophy and Social Science Edition)] 6: 105–108.

Zhang, Li. 2002. *Strangers in the City: Reconfigurations of Space, Power, and Social Networks within China's Floating Population.* Stanford, CA: Stanford University Press.

Zhang, Qunhui 张群辉. 2015. "Zhengzhi shengtaixue de guoji jingyan yu zhongguo shijian" 政治生态学的国际经验与中国实践 [International Experience of Political Ecology and China Practice]. *Guowai shehui kexue* 国外社会科学 [Social Sciences Abroad] 3: 23–29.

Zhang, Rui-Hang 张瑞杭, Qi-Jun Liu 刘启俊, and Guang-Jian Huang 黄广健. 1996. "Dongjiang xiayou heduan hechuang xiajiang yinfa de wenti he duice tanjiu" 东江下游河段河床下降引发的问题和对策探讨 [Problems Caused by the Deepening of the Riverbed in the Lower Reaches of the Dongjiang River]. *Redai dili* 热带地理 [Tropical Geography] 16(3): 236–243.

Zhang, Sheng-Zhen 张声震. 1997. *Zhuangzu tongshi* 壮族通史 [General History of the Zhuang]. Beijing: Publishing House of Minority Nationalities.

Zhang, Yinfeng 张银锋. 2013. *Cunzhuang quanwei yu jiti zhidu de yanxu: "Mingxingcun" gean yanjiu* 村庄权威与集体制度的延续: "明星村"个案研究 [The Authority of the

Village and Continuance of the Commune System: A Case Study of Star Village]. Beijing: China Social Science Press.

Zhao, Yang 赵杨. 2014. "Fandu chuji zhanduan bufa guanshang 'daocai tongmeng'" 反渎出击斩断不法官商'盗采同盟' [Anti-Blasphemy Attack Cuts Off the Judge and Business "Pirate Procurement Alliance"]. *Southern Net.* October 17. http://epaper.southcn.com/nfdaily/html/2014-10/17/node_15.htm (accessed December 30, 2017).

Zhou, Dongtao 邹东涛, and Rihui Ouyang 欧阳日辉. 2008. *Zhongguo suoyouzhi gaige Sanshinian: 1978–2008* 中国所有制改革30年: 1978–2008 [Thirty Years of Ownership Reform in China: 1978–2008]. Beijing: Social Sciences Academic Press.

Zhou, Feizhou 周飞舟. 2012. "Caizheng zijin de zhuanxianghua jiqi wenti jianlun 'xiangmu zhiguo'" 财政资金的专项化及其问题兼论 "项目治国" [The Problems of Fiscal Earmarked Funds: On "Governing the State through Programs"]. *Shehui* 社会 [Society] 1: 1–37.

Zhou, Xueguang 周雪光. 2011. "Quanwei tizhi yu youxiao zhili: dangdai zhongguo guojia zhili de zhidu luoji" 权威体制与有效治理:当代中国国家治理的制度逻辑 [Authoritative System and Effective Governance: Institutional Logic of Contemporary China's National Governance]. *Kaifang shidai*开放时代 [Open Times] 10: 67–85.

Zhu, Kai-Bin 朱凯斌, Yu-Sheng Yang 杨玉生, Xiao-Sheng Liu 刘小生, and Lin-Juan Yuan 袁林娟. 2015. "Jiyu butong gongcheng fenlei fangfa de shatu dingming ji xiangying wuli lixue canshu duibi fenxi" 基于不同工程分类方法的砂土定名及相应物理力学参数对比分析 [Sand Naming and Comparative Analysis of the Corresponding Physical and Mechanical Parameters using Different Engineering Classification Methods]. *Dizhen gongcheng xuebao* 地震工程学报 [China Earthquake Engineering Journal] 37 (S1): 17–21.

Zhu, Qian 朱倩. and Yan-jia Qin 覃延佳. 2016. "Chaoyue jingji lixing: laodongli liudong beijing xia de sangzang yishi yu cunli zhixu chongsu" 超越"经济理性"：劳动力流动背景下的丧葬仪式与村里秩序重塑 [Beyond 'economic rationality' Funeral Ritual and Remodeling of Village Order in the Background of Labor Mobility: A case study of Mei Village]. *Xinan bianjiang minzu yanjiu* 西南边疆民族研究 [Southwest Frontier Ethnic studies] 21: 150–156.

Zhu, Qian 朱倩. 2018. "Gonggong ziyuan zhili kunjing: shilun guangxi cunzhuang zai hesha shichanghua guochengzhong de jueli" 公共资源治理困境：试论广西村庄在河砂市场化过程中的角力 [The Dilemma of Public Resource Governance: The Struggle of Villages in the Process of Marketization of River Sand]. *Guangdong shehui kexue* 广东社会科学 [Soil & Social Sciences in Guangdong] 5: 210–220.

Zhu, Qian 朱倩. 2019. "Shichanghua beijingxia de xiangcun gonggong ziyuan guanli kunjing ji chongtu tiaojie" 市场化背景下的乡村公共资源管理困局及冲突调节 [The Dilemmas of Public Resource Management and Conflict Mediation under a Marketization Economy]. *Shengtai jingji pinglun* 生态经济评论 [Ecological Economic Review] 9: 47–57.

Zhu, Qian. 2020. "River Sand as a Disputed Resource: A Case of Illegal Sand Mining Near Zhuang Villages in Southwest China." PhD diss., Department of Cultural and Social Anthropology, University of Cologne, Germany.

Zhu, Rong-Ji 朱镕基. 2001. "Changjiang hedao caisha guangli tiaoli" 长江河道采砂管理条例 [The Regulations on Sand Mining Management of the Yangtze River]. Decree No. 320 of the State Council of the People's Republic of China [国务院令第320号]. Beijing: The State Council of the People's Republic of China [中国国务院].

Zhu, Ying 朱英. 2016. "Zhonggong zhongyang guowuyuan bangongting yinfa 'Guanyu quanmain tuixing hezhangzhi de yijian'" 中共中央办公厅国务院办公厅印发《关于全面推行河长制的意见》 [The General Office of the CCP Central Committee General Office of the State Council Issued the "Opinions on Comprehensively Implementing the River Chief System"]. http://www.gov.cn/xinwen/2016-12/11/content_5146628.htm (accessed December 20, 2016).

Index